THE GOLDEN YEARS
OF THE CLYDE STEAMERS
(1889-1914)

THE
GOLDEN YEARS
OF THE
CLYDE STEAMERS
(1889-1914)

ALAN J. S. PATERSON

DAVID & CHARLES : NEWTON ABBOT

ISBN 0 7153 4290 8

First published 1969
Second impression 1979

Printed in Great Britain by
Redwood Burn Ltd Trowbridge & Esher
for David & Charles (Publishers) Limited
Brunel House Forde Road
Newton Abbot Devon

CONTENTS

		Page
	LIST OF ILLUSTRATIONS	7
	INTRODUCTION	11

Chapter

1 THE CLYDE IN THE LATER EIGHTIES — 15
Early years . railway failures . tourist boats . railway steamers . private owners . the Campbells of Kilmun

2 THE RISE OF THE CALEDONIAN (1889-1894) — 34
Steamer plans . a Parliamentary rebuff . the Caledonian Steam Packet Company, Ltd . new steamers . Ardrossan and Arran . Wemyss Bay improvements . new engines and boilers

3 SOUTH WESTERN RESURGENCE (1890-1894) — 67
Parliamentary battles . steamers old and new . railway problems . the new Prince's Pier . racing days

4 UP RIVER AND DOWN FIRTH (1889-1894) — 95
The Craigendoran route . Wemyss Bay finale . the *Lord of the Isles* . new steamers at Craigendoran . a new *Lord of the Isles* . Buchanan changes . return of the *Victoria* . Campbeltown routes . the Broomielaw at the Fair

5 VICTORIAN CLIMAX (1895-1900) — 118
Craigendoran additions . Clydebank steamers for 'Caley' and 'Sou' West' . more new vessels . Sunday breaking . water tube boilers . a North British compound . the wreck of the *Redgauntlet*

6 THE NEW CENTURY—I : THE TURBINE ADVENTURE (1901-1912) — 152
The Parsons Turbine . the *Viper* and *Cobra* . the turbine syndicate . the *King Edward* and *Queen Alexandra* . a 'Caley' turbine . the *Atalanta* . fire on board . a new *Queen Alexandra*

7 THE NEW CENTURY—II (1901-1914) — 183
The Loch Eck route . Caledonian changes . the *Marmion* . the siege of Millport . the loss of the Kintyre . railway retrenchment . Broomielaw revival

Chapter		*Page*
8	A CLYDE MISCELLANY	212
	Racing risks . the Clyde sewage problem . the smoke fiend . fog and storms . yachting . a Campbell visitor . music on board . pier dues	
9	BOILERS AND MACHINERY	234
	Single diagonals and haystacks . compounds and Navy boilers . triple expansion . oscillating engines and 'steeples' . paddle wheels . points of detail	

APPENDICES

1	Colours and Uniforms	251
2	The Clyde Steamer Fleet in 1914	258
3	Clyde Piers and Ferries, 1889 to 1914	259
4	P.S. *Neptune* Cruises, July 1896	261
5	The Turbine Syndicate Agreement, 1901	262
6	Some Clyde Holiday Advertisements, June 1898	265
7	Specification of P.S. *Lady Rowena*, 1891	266
8	Memorandum as to Campbeltown Traffic, June 1899	270
9	Caledonian Steam Packet Company, Ltd : Financial Accounts, 1892 and 1893	272
10	Fleet Lists	276
	AUTHOR'S NOTES AND ACKNOWLEDGMENTS	287
	BIBLIOGRAPHY	289
	INDEX	291

LIST OF ILLUSTRATIONS

LIST OF PLATES

Page

The *Columba* arriving at Rothesay *(G. E. Langmuir collection)*　17

The *Davaar* in original condition *(G. M. Stromier & L. J. Vogt collection)*　18

The *Grenadier*, as altered in 1903 *(G. M. Stromier & L. J. Vogt collection)*　18

The *Duchess of Rothesay* at Corrie Ferry *(G. E. Langmuir collection)*　35

The temperance steamer *Ivanhoe (Adamson/Robertson collections, by courtesy of K. V. Norrish)*　36

The Sabbath-breaking *Victoria (Adamson/Robertson collections, by courtesy of K. V. Norrish)*　36

The second *Lord of the Isles* arriving at Rothesay *(G. E. Langmuir collection)*　53

The *Neptune* at Ayr *(G. M. Stromier & L. J. Vogt collection)*　54

Buchanan's *Isle of Arran (Adamson/Robertson collections, by courtesy of K. V. Norrish)*　54

The *Lucy Ashton* in the mid-nineties *(G. E. Langmuir collection)*　71

The *Lady Rowena* on the Arrochar service *(Adamson/ Robertson collections, by courtesy of K. V. Norrish)*　71

The *Marchioness of Lorne* approaching Dunoon *(G. M. Stromier & L. J. Vogt collection)*　72

The *Viceroy* arriving at Cove *(G. M. Stromier & L. J. Vogt collection)*　72

The *Jupiter* in Rothesay Bay *(Adamson/Robertson collections, by courtesy of K. V. Norrish)*　89

The *King Edward* on trials off Skelmorlie *(National Maritime Museum)*　90

7

Page

The first *Queen Alexandra (National Maritime Museum)*   107

Original propeller arrangement of the *King Edward (National Maritime Museum)*   107

The *Duchess of Argyll* in modified condition *(Adamson/ Robertson collections, by courtesy of K. V. Norrish)*   108

The *Atalanta* passing Gourock *(Adamson/ Robertson collections, by courtesy of K. V. Norrish)*   108

The *Marmion* in 1906 *(G. M. Stromier & L. J. Vogt collection)*   125

The *Strathmore* leaving Dunoon *(G. E. Langmuir collection)*   126

The *Eagle III* sailing down river *(G. E. Langmuir collection)*   126

The second *Lord of the Isles* at Inveraray *(G. E. Langmuir collection)*   143

The first *Lord of the Isles* and the *Adela* at Rothesay in the eighties *(G. E. Langmuir collection)*   144

The tandem compound machinery of the *Caledonia (Engineering)*   161

The single diagonal engine of the *Madge Wildfire (G. M. Stromier & L. J. Vogt collection)*   162

The compound engine of the *Neptune (G. M. Stromier & L. J. Vogt collection)*   162

The compound engine of the *Jupiter (G. M. Stromier & L. J. Vogt collection)*   179

The tandem triple engine of the *Duchess of Montrose (G. M. Stromier & L. J. Vogt collection)*   179

The *Viceroy* and other steamers in Rothesay Bay *(G. E. Langmuir collection)*   180

Captain Angus Campbell of the *Columba (G. M. Stromier & L. J. Vogt collection)*   197

Captain A. McKellar and the crew of the *Galatea (G. M. Stromier & L. J. Vogt collection)*   197

Lord Dunraven's racing cutter *Valkyrie (Author's collection)*   198

The *Valkyrie III* racing off Gourock *(Author's collection)*   198

The veteran *Iona* in Rothesay Bay *(G. E. Langmuir collection)*   215

P. & A. Campbell's *Britannia* on trials on the Clyde *(G. E. Langmuir collection)*   215

Page

Three paddleboxes: *Talisman (G. M. Stromier & L. J. Vogt collection); Duchess of Fife (G. M. Stromier & L. J. Vogt collection); Queen Empress (G. M. Stromier & L. J. Vogt collection)*   216

LIST OF DRAWINGS

Map of the River and Firth of Clyde	10
Destination boards	33
Fireworks at Wemyss Bay—a Caledonian extravaganza of 1890	62
Turbine destination board	182
Engine room plate, *Duchess of Fife*	211
Dawson Reid's cruise advertisements of 1901	230
A typical haystack boiler	235
Navy boiler	238
Double-ended boiler	240
A standard 7-float Caledonian paddle wheel	246
South Western paddlebox crest	250
Pennants	284-5
Funnels	286

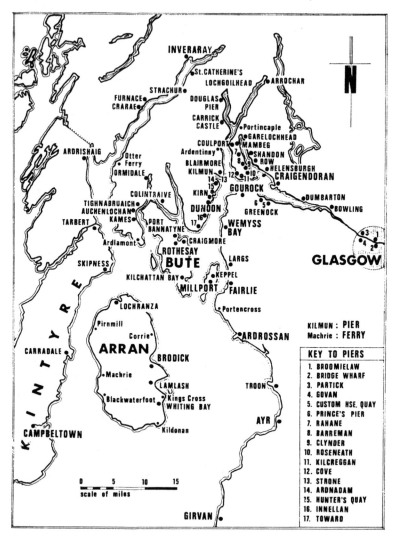

Map of the River and Firth of Clyde

INTRODUCTION:
THE RIVER AND FIRTH OF CLYDE

NO other coastal waters around the British Isles are quite like those of the Firth of Clyde. Covering an area of some 1,200 square miles, they form a vast estuary sheltered from the outer waters of the Atlantic Ocean by the long peninsula of Kintyre, and the inner reaches have long been recognised as a safe anchorage for the largest seagoing ships. The southern shores are lowland in character, and here are found the principal towns such as Greenock, Port Glasgow, Largs and Ayr. The northern and western shores, on the other hand, have more in common with the Scottish Highlands, being generally mountainous and uncultivated, while long sea lochs cut far into the hills. Towns are fewer and smaller, and the atmosphere of Inveraray, Tarbert and Campbeltown has little in common with that of the larger communities in Renfrewshire and Ayrshire. The firth is given much character by islands, amongst which the mountainous bulk of Arran dominates the scenery of the outer reaches. The beauty of Bute is much less dramatic, but this more pastoral island, together with the two Cumbraes close to the Ayrshire coast, has great charm.

Scenically the firth covers a wide spectrum, from the industrial background of the upper river to the remoteness of the head of Loch Fyne, its longest arm, and from the bustling, holiday atmosphere of Dunoon and Rothesay to the peace of Arran villages. The narrow recesses of the fjord-like Loch Long contrast sharply with the spacious waters of the lower firth, and the gloom of Loch Striven with the charms of the Holy Loch. There is a wide variety in a relatively small area, and therein lies the enduring attraction of this famous estuary.

Glasgow has been the commercial capital of the area ever since a tremendous expansion of trade with the American colonies followed the Union of the Scottish and English Parliaments in 1707. Trade in tobacco and cotton carried the city to prosperity and paved the way to huge expansion during the Industrial Revolution. The Broomielaw, the harbour in the heart of Glasgow, well up-river from normal navigable limits, was enlarged during the later eighteenth century and the River Clyde itself widened and deepened into a tidal ship canal allowing ships to come into the city itself to unload.

The attractions of the Clyde estuary were quickly appreciated as Glasgow's importance grew and there are records of an embryo of summer traffic from the city in the earliest years of the nineteenth century. Nevertheless, transport was slow and uncertain and it was not until the advent of steam power that the Clyde area as a whole began to be developed. Progress thereafter was rapid and in a very short time the steamboat was an accepted sight on all parts of the firth.

This is the story of twenty-five years in the history of the Clyde passenger steamer. Its origins were ancient, going back as far as the Napoleonic Wars when, in 1812, Henry Bell inaugurated the first commercially successful service by steam vessel on the waters of the river and firth with the *Comet*, which plied between Glasgow and the fashionable watering place of Helensburgh. Development came rapidly and by the twenties there was a fleet of steamboats sailing to many coastal resorts. Circumstances combined to encourage further expansion, for geographical factors prevented railways from becoming an effective competitive force, but the Clyde itself was easily navigable right into the heart of Glasgow, and private steamboat owners flourished on the burgeoning traffic from the city to towns such as Dunoon, Rothesay and Millport, besides a hundred smaller villages and hamlets. By mid-century the teeming population of Glasgow, a city thrusting and expanding into industrial prominence, had discovered in the Clyde coast a natural holiday area; fares were cheap, distances short, and the town-dweller grateful for the opportunity of relaxing in peaceful surroundings practically on his doorstep. The cult of the day trip 'doon the watter' became firmly established and with it the affection of the Glaswegian for 'the boats' which took him and his family away, if only for brief interludes, from the grime, poverty and overcrowding of a Victorian industrial city on the make.

By the eighties, steamers were an integral part of the life of the community, but signs of impending change could be detected. Many private owners of the early years had died, or were old men; some had given up business either by choice or necessity under the pressure of changing economic circumstances. Larger fleets were more common and it became evident that the old 'single steamer' owner had no future. Gradually the railway companies running to the coast began to contemplate extensions and competition flared. Parliament frowned on railway owner-ship of steamboats, but means were found of avoiding its pro-hibition and as the eighties drew to a close the scene changed radically as these large companies threw themselves into all-out efforts to capture the coast traffic. The following quarter of a century, ending in the outbreak of the Great War, was the most colourful and extravagant period of Clyde steamer history. The following chapters trace the story from the last brilliant years of Queen Victoria's reign through the Edwardian mid-summer and on to August 1914, when the old order was finally swept away.

CHAPTER ONE

THE CLYDE IN THE LATER EIGHTIES

STEAMBOAT traffic developed swiftly in the early years of the nineteenth century, for the geographical features of the Clyde area were peculiarly favourable to the new form of transport. Railway promoters, on the other hand, found it uncongenial territory and the first lines were projected simply to link Glasgow with coastal towns in the counties of Dunbarton, Renfrew and Ayr. Early attempts by the railways to run steamboats on their own account met with hostility from the established shipowners and were generally unsuccessful. The 1860s and 1870s saw the heyday of the private steamer owner, but as the century advanced the tendency was towards larger fleets and several owners went out of business. A corresponding tendency to amalgamations amongst Scottish railways eventually brought three large companies into competition for the Clyde coast traffic, and during the 1880s it became evident that the influence of the private shipowner was in decline, although at the end of the decade the Clyde steamer fleet, taken as a whole, was unsurpassed in size and quality of service in British coastal waters.

EARLY YEARS

For many years now it has been natural to think of a journey to a Clyde coast resort as involving train as well as steamer, transferring from one to the other at one of the railheads on the firth. It was not always so. There was a time when the principal route to the lower reaches was that of the steamboats which sailed all the way to their destinations from the Broomielaw, the harbour in the centre of Glasgow. From the earliest days of steamboat history, vessels plied to coast towns and villages from the

15

city, establishing and consolidating a trade which was nearly thirty years old when the first railway appeared to challenge their position. By the middle of the 1830s more than three dozen steamships were recorded as sailing to places within the Clyde estuary as far afield as Stranraer, Ayr, Campbeltown, Arrochar and Lochgilphead. Prospects for speculators were tempting, and a large fleet rapidly built up as new owners entered the field.

RAILWAY FAILURES

The Glasgow, Paisley and Greenock Railway was opened in 1841, running from Bridge Street station, Glasgow, to Paisley, Langbank and Greenock (Cathcart Street). The Glasgow terminus, long since swept away, occupied the site of the extensive sidings immediately to the south of the present Central station while the Greenock station was approximately in the same position as today's Greenock Central. The G.P. & G. terminus was situated close to Custom House Quay where many of the up-river steamers called in passing, and it was not long before the new railway entered into an arrangement with the Bute Steam Packet Company under which its vessels *Isle of Bute* and *Maid of Bute* were advertised to sail in connection with the trains. The relationship became closer in 1844 when the railway company assumed direct control of the steamboat concern and purchased three additional steamers, with which an improved service was operated. The advantages of the rail and steamer route seemed obvious, but the venture met with powerful opposition from the private steamboat owners operating from Glasgow. Losses were incurred, and in 1846 the steamer interests were sold, arrangements being made with private firms for the maintenance of connections at Greenock. The reasons for the failure of this, the first combined rail and steamer venture of its kind in Britain, are not difficult to appreciate. Lack of experience on the part of the railway company was fatal in the face of determined opposition from the steamboat owners, uninhibited in those years by speed restrictions on the upper reaches of the river. In the event the superior speed

Page 17 : MacBrayne majesty : the *Columba* arriving at Rothesay.

Page 18: Clipper bows: *(above)* the *Davaar* as originally built; *(below)* the *Grenadier* as altered in 1903. Both vessels were photographed on the upper river, near Glasgow.

of train travel was offset by the disadvantage of having to walk
between station and quay at Greenock. Custom House Lane,
which passengers had to traverse, was an unpleasant thorough-
fare, particularly in wet weather, and the majority of travellers
preferred to sail directly from Glasgow rather than suffer inconvenience. But perhaps the most powerful incentive was the attraction of cheaper fares by steamer and in the end it was hardly
surprising that the Greenock Railway's venture failed. In 1847,
however, the company was absorbed by the Caledonian Railway
and the new proprietors attempted to re-establish a steamboat
connection in the early 1850s, but the pattern was repeated and
again the vessels were disposed of, leaving the traffic in the hands
of private operators.

Nearly two decades passed before a railway company again
attempted to run steamers on its own account. The Greenock &
Wemyss Bay Railway Company was opened on 15 May 1865,
running from a junction with the Caledonian Railway at Port
Glasgow and thence over the Renfrewshire hills via Inverkip to
its terminus on the coast. An associated concern, the Wemyss Bay
Steamboat Company Limited, was formed for the purpose of
running steamers in association with the railway. The proprietors
evidently sensed that circumstances were more favourable and
set out with vigour to capture the trade between Glasgow and
Rothesay and Millport. It seemed that the new venture could
scarcely fail. Wemyss Bay was the nearest mainland port to
Rothesay to which a railway service could reasonably be run, and
as far as speed was concerned the advantage lay heavily with
the railway route. Realising the threat to their interests, the steamboat proprietors opposed the Greenock & Wemyss Bay Railway
Bill as far as the House of Lords but were finally forced to admit
defeat. They were particularly vulnerable at this period, for the
Clyde had been stripped of its newest and fastest steamers during
the early 1860s by agents of the southern Confederacy, desperate
to acquire good ships to run the coastal blockade imposed by the
Federal Government during the American Civil War. In consequence, vessels remaining on the firth were generally old and

slow, and it was confidently anticipated that the new Wemyss Bay Steamers would outpace the outdated fleet then maintaining services on the lower reaches.

The Wemyss Bay managers ordered four steamers and prepared an ambitious timetable for the summer of 1865, but a combination of indifferent management and ill-fortune ruined their prospects. Collectively the new vessels were disappointing in the all-important matter of speed, and it was embarrassing when they were found unable to match their older rivals. Worse, however, was their inability to keep to the timetables, a fault which bedevilled operations during the first summer. Despite subsequent efforts to remedy the situation the managers of the Wemyss Bay Steamboat Company were unable to prevent its eventual collapse, and it was reported to the railway shareholders in September 1869 that the steamers had been withdrawn. Other steamer operators had temporarily given connecting services but these were not regarded as satisfactory until a Captain Gillies agreed to conduct the whole service from Wemyss Bay, an arrangement that continued for over twenty years during which he was joined, and ultimately succeeded, by his son-in-law Captain Alexander Campbell.

The next to enter the arena was the North British Railway which, in 1866, inaugurated an ambitious programme of sailings from Helensburgh, on the north shore of the firth, to Ardrishaig. Two large, lavishly appointed steamers were built and managed through an intermediary, the North British Steam Packet Company. It too was dogged by ill-luck; one of the new vessels suffered a major breakdown during the first season and the company had to resort to expensive chartering to maintain the service. Worse was to follow, for in the same year a financial scandal seriously damaged the credit and reputation of the parent company, whose chairman and some officers were found to have been criminally mismanaging its affairs to such an extent that it was to all intents and purposes bankrupt. Retrenchment was inevitable and the Steam Packet Company was the first to suffer. Services were drastically curtailed and one of the steamers sold, leaving the other to ply only as far as Dunoon and the Holy Loch, a state of affairs that prevailed until 1883 when it was found possible to expand again on a sounder footing. Once more a railway company had failed to undermine the position of the private owners but in this instance at least the North British company succeeded in establishing a permanent foothold on the Clyde and for this reason it must be accorded the honour of being the oldest continuously-operated rail and steamer service from Glasgow to the coast.

In 1869 the Greenock & Ayrshire Railway was opened between Elderslie, Kilmalcolm[1] and Greenock, where a new railhead, Prince's Pier, was built. The line was worked by the Glasgow & South Western Railway, which absorbed it in 1872. This was a much harder route to work than the Caledonian line to Greenock; climbing steadily through Bridge of Weir and Kilmalcolm, it reached a summit high above Greenock before descending steeply to the terminus. In later years it was to be the scene of herculean performances by the little Smellie and Manson express engines of the South Western Railway. In spite of its operating problems the new railway was much superior to the Caledonian route in being further down river and the trains ran practically on to the pier, so that it was simply a matter of time before it drew traffic from its rival. The experience of other railways may well have been in the minds of the Greenock & Ayrshire and Glasgow & South Western directors, for no attempt was made to run steamers from Prince's Pier. Arrangements were made instead with Captain Alexander Williamson to place his steamers on the railway connections, and these were continued amicably for over twenty years.

The position at the start of the 1870s can be summarised briefly as follows. The bulk of river traffic went from Broomielaw, being conducted by private owners. The North British Steam Packet Company operated a service to Dunoon and Holy Loch piers. Williamson's fleet maintained a service from Prince's Pier to Rothesay and Port Bannatyne, gradually taking traffic from the

[1] Now Kilmacolm.

Caledonian route via Cathcart Street, Greenock, and Custom House Quay. At Wemyss Bay, Gillies and Campbell operated services to Port Bannatyne, Rothesay and Millport, where they enjoyed a virtual monopoly. Arran traffic went by the Glasgow & South Western route to Ardrossan and thence in privately owned steamers to Brodick and Lamlash. The pattern established remained for the better part of the ensuing twenty years.

Following a period of consolidation after the amalgamations and adjustments of the mid-Victorian era, the 1880s saw a fresh surge of expansion by the Scottish railways and in course of time competition flared as their interests inevitably conflicted. This was the decade of the great Tay and Forth Bridges and of the Caledonian thrust to Oban, highlights in the general improvement and modernisation of facilities and equipment by all companies as they entered the period of their greatest prestige and influence.

The Caledonian Railway directors were discontented with their company's inferior position in relation to the coast traffic through Greenock; as early as 1869 land had been acquired on the foreshore of Gourock Bay and powers taken to extend the railway there, but these were allowed to lapse. Fifteen years later, however, a fresh scheme was prepared and approved by Parliament in the face of strong opposition; this provided for the extension of the Greenock railway to Gourock, a distance of three and a half miles through and under Greenock in cuttings and tunnels, at an eventual cost of £600,000. Emerging from the last tunnel at Fort Matilda, the new line was planned to sweep down to a terminus on the west side of Gourock Bay, where the largest and finest pier on the Firth of Clyde was to be constructed. The new terminus was built on an embankment half a mile long formed by the spoil excavated from the tunnels and cuttings in Greenock, and the extensive steamer berths gave accommodation for several vessels at one time. It was intended to build up goods traffic and the average depth of water at low tide was twenty-two feet to allow large ships to berth and discharge. There was generous siding accommodation, while the passenger station was large and imposing, even though the adoption of a mock-Tudor style of architecture introduced a bizarre note.

As the new Caledonian railway to Gourock took shape, the Clyde steamer fleet entered the season of 1888, the last in which private owners predominated on the Clyde. Some forty river steamers were in service, operated by a dozen owners, most of whom were swept out of business during the nineties, and in order that the developments of that decade may be fully understood it is necessary to review the position at the start of the period.

TOURIST BOATS

Foremost of all the steamers on the river was David MacBrayne's mighty *Columba*, acknowledged as supreme in a small group of 'tourist' steamers which plied only in summer, catering for the West Highland landowners and their guests bound for estates in the north west. It is essential to remember the wide class distinctions of Victorian days in considering the *Columba*—she was, above all, the ship in which the *best people* sailed, and this was her reputation throughout the quarter of a century before the Great War levelled things down for good. She was built in 1878 by J. & G. Thomson, of Clydebank, and even when new was old-fashioned in some respects. Her curved bow and square stern were of the sixties, but with two large funnels fore and aft of the paddleboxes she was an imposing vessel. She was driven by the distinctly old-fashioned type of oscillating machinery commonly used twenty years earlier, and even by the nineties was mechanically outdated. There, however, the conservatism of her owner ceased, for in all other respects the *Columba* was equipped in a manner that befitted the MacBrayne flagship, sailing on the most important route served by the fleet. The *Columba* sailed daily from Glasgow to Greenock, and on by way of Dunoon and Rothesay through the Kyles of Bute to Tarbert and Ardrishaig, carrying passengers for Oban and the north. She entered service at a time when competition from the rival *Lord of the Isles* threatened the position on the Clyde, and the imminent completion of the Callander & Oban Railway offered a more serious threat to through traffic. The *Columba* was therefore designed to offer passengers unprecedented luxury, an aim in which her builders amply succeeded. The first vessel on the Clyde to have her deck saloons built out to the full width of the hull, advantage was taken of this to furnish the ship lavishly, while additional features such as a post office and a barber's shop were also incorporated. With such a magnificent vessel in service the double threat to MacBrayne's traffic was successfully contained and his position secured for many years.

When traffic on the Ardrishaig station was lighter during the late spring and early autumn months, the *Columba* was replaced by her older consort, the *Iona*, built in 1864. By Clyde standards she was definitely old-fashioned by the later eighties, having the same type of machinery as the *Columba*, and having narrow deck saloons. Nevertheless, she too was kept in immaculate condition and the travelling public seemed to regard her as ageless. Although much smaller than the *Columba*, she resembled her in general appearance, although there were numerous features by which the two vessels could readily be identified. In the summer months the *Iona* augmented the service between Ardrishaig and Glasgow.

During the winter the Ardrishaig mail service was maintained by the paddle steamer *Grenadier*, built in 1885. She was an attractive vessel with a clipper bow and two funnels; these were quite close together, spoiling her outline, but after reboilering at the turn of the century the defect was remedied and the *Grenadier* came into her own as one of the handsomest ships on the Clyde. As in the case of the *Iona* and *Columba*, she was driven by oscillating machinery, the last set ever to be made for a Clyde steamer but also the first—and only—set of engines of this type working on the compound system. Her deck saloons were made the full width of the hull and generally the *Grenadier* was well fitted out, although not on the same luxurious scale as the *Columba*. Her task, which she carried out successfully for a long period, was to maintain the winter mail run when sparseness of traffic demanded economical working; during the summer her usual station was the tourist route round Mull to Staffa and Iona, sailing out of Oban.

Second only to the *Columba* in importance and reputation was the *Lord of the Isles*, owned by the Glasgow & Inveraray Steamboat Company, Limited. Built in the year before her rival entered service, the *Lord of the Isles* introduced a new standard of luxury to the Clyde fleet, far beyond anything offered even by the *Iona*. The new steamer was placed on the long, 180-mile daily return sailing from Glasgow to Inveraray, at the head of Loch Fyne, and rapidly became a favourite on this station. Her clientele was much the same as that of the *Iona* and she duplicated most of that vessel's ports of call, causing the latter's withdrawal in favour of the *Columba*. As befitted a first class tourist steamer, the *Lord of the Isles* was richly furnished and her catering standards were high. She was an attractive vessel, beautifully proportioned, and her striking colours—red, white and black funnels—did much to enhance her already elegant appearance.

The *Lord of the Isles* was closely associated with the steamers belonging to the Lochgoil & Lochlong Steamboat Company, Limited which shared the same funnel colours and also operated under the managership of Malcolm T. Clark. None of these vessels, however, was in the same high class as the Inveraray steamer, and all were much smaller. The oldest was the *Windsor Castle*, of 1875, which sailed to Lochgoilhead. She had a long, narrow deck saloon aft, but none forward. Like many of her contemporaries she was fitted with a single cylinder, diagonal engine and steam was supplied by the almost universal haystack boiler. Four years her junior was the *Edinburgh Castle*, of much the same general design, but distinguished by the largest paddle wheels of any Clyde steamer before or since. Under the command of Captain William Barr, for many years the doyen of Clyde skippers, this vessel was long a favourite on the Lochgoilhead route. The third steamer belonging to the company was the *Chancellor*, a smart little vessel acquired second-hand in 1885 from the Lochlong & Lochlomond

Steamboat Company. Built in 1880, she had spent her entire career on the route between Helensburgh (Craigendoran after 1882) and Arrochar, in connection with Loch Lomond steamers. Of more modern appearance than the other steamers, the *Chancellor* had narrow deck saloons fore and aft and was driven by a set of simple expansion, two cylinder diagonal machinery.

Amongst the highest class of excursion steamers, the *Ivanhoe* was unique in being operated in accordance with temperance principles. She had been built by D. & W. Henderson in 1880 and in general design, both of hull and engines, resembled the *Lord of the Isles*. She was owned by the Frith of Clyde Steam Packet Company Limited, this being the trading title of a syndicate of Clyde shipowners who were prepared to give practical support to the view that a Clyde passenger steamer could be run successfully without the aid of 'the domen drink'. The *Ivanhoe* entered service when the social problem of drunkenness was acute in Glasgow and all too often passengers on board steamers suffered embarrassment from fellow travellers under the influence of what were then referred to by the genteel as 'ardent spirits'.

The *Ivanhoe*'s appeal was to those who wished to enjoy a day's outing guaranteed free of the drunken excesses so frequently encountered on other vessels. The route which she made her own was from Helensburgh—and, for several seasons, Craigendoran—to Greenock, Kirn, Dunoon, Wemyss Bay and Rothesay, and thence through the Kyles of Bute to the Arran ports of Corrie, Brodick, Lamlash, King's Cross and Whiting Bay. Gourock was added to the ports of call after 1888. Scenically this was one of the finest routes on the firth and the grandeur of the north Arran mountains did much to popularise the *Ivanhoe*'s sailings. Most of the credit, however, belonged to one man, Captain James Williamson, son of Alexander Williamson of the Prince's Pier steamers. James Williamson, trained both as engineer and skipper, was one of the most competent masters on the firth, and destined to play a greater part than any other man in the shaping of Clyde steamer services during our period. He was secretary and

manager of the Frith of Clyde Steam Packet Company and
master of the *Ivanhoe* for several years. His was the initiative
in turning the steamer out in a manner more in keeping with a
private yacht. The vessel herself was kept immaculately clean,
the crew wore smart, naval-style uniforms with white sailor
jerseys, and discipline was strict. The *Ivanhoe* was a graceful
steamer, painted in attractive colours, and she was well run; the
public rose to her, and success was instant. A curious thing
became apparent, as James Williamson noted later : the owners
had relied on the strong support of temperance organisations,
but in fact it was the patronage of 'the moderate class of the
community' which did most to ensure that the venture succeeded.
Disgusted by riotous scenes on other steamers, many people
responded enthusiastically to the smart, clean 'teetotal boat' on
which wives and children could safely be taken for a day's outing in comfort and peace. The *Ivanhoe* therefore had a special
place in the affections of Glaswegians during the eighties and her
appeal was diminished only in the mid-nineties when new railway steamers appeared to challenge her high standards.

RAILWAY STEAMERS

The only railway fleet on the river in 1888 was that of the North
British Steam Packet Company, whose operating headquarters
were at Craigendoran. This was not a limited liability company,
but a co-partnership of directors and senior officers of the North
British Railway. Unlike any other Clyde fleet, the North British
was controlled ultimately from Edinburgh, but the appointment
of the able and energetic Robert Darling as secretary of the company in 1882 ensured that it would be well run. Local control
was strengthened in 1893 when Darling and his steamboat
officials were transferred to Glasgow, and he himself went to live
in Helensburgh until his death, at the early age of fifty-two, in
1899. The Craigendoran fleet in 1888 totalled five, of which the
newest vessel was the saloon steamer *Lucy Ashton*, built in that
year by T. Seath & Co of Rutherglen and engined by Hutson &

Corbett with a conventional single diagonal set of machinery, steam being supplied by a haystack boiler. She had been built for the Holy Loch service, which she operated with the *Diana Vernon*, three years her senior, and of similar design. The latter, however, was often employed on the Gareloch station as well, her consort there being the oldest member of the fleet, the *Gareloch*, built in 1872. The *Gareloch* was of the so-called 'raised quarter deck' type, much in vogue during the seventies and eighties. Vessels of this design were built with their quarter decks at main rail level, allowing better lighting and ventilation in the main saloon below as compared with steamers of the older flush deck type.

The remaining North British steamers, the *Jeanie Deans* and *Guy Mannering*, were both of the raised quarter deck design, of which each was considered an outstanding example. The *Guy Mannering* had been built by Caird & Co of Greenock in 1877 as the *Sheila*, for Gillies and Campbell's Wemyss Bay fleet, in which she had established a reputation as an unusually fast ship. The North British Steam Packet Company acquired her for the extension of services following the opening of Craigendoran Pier in 1883. She was joined by the *Meg Merrilies* in the same year but the *Meg* was a luckless vessel from the start, destined for the most chequered career of any Clyde steamer, and the North British returned her to the builders, Barclay, Curle & Co, at the end of the season. She was succeeded by the *Jeanie Deans* from the same yard in 1884. This was a splendid ship, fast and reliable, and in company with the *Guy Mannering* she became well-known and popular on the Rothesay route. Built to retrieve her builders' reputation after the *Meg Merrilies* fiasco, the *Jeanie Deans* reverted to the well-tried raised quarter deck arrangement with single engine and haystack boiler. She had the distinction of being the last raised quarter deck vessel to be built for Clyde service, all subsequent steamers having full deck saloons.

Although privately owned, the steamers operated by Captain Alexander Campbell in connection with the Caledonian Railway services to Wemyss Bay were to all intents and purposes railway boats, sailing almost exclusively on regular service runs to Rothesay and Port Bannatyne, Innellan, Largs, Millport and Kilchattan Bay. The largest and most modern vessel was the *Victoria*, built by Blackwood & Gordon in 1886. With full deck saloons fore and aft, and propelled by double diagonal engines, she was well suited for excursion work, but her owner's insistence on employing her on such duties instead of railway connections led ultimately to a break with the railway company and the withdrawal of the *Victoria* and her consorts from the Clyde. She returned, under different management, during the nineties, and we shall have occasion to encounter her again in these pages. The other three Wemyss Bay steamers were much smaller and older vessels, having been built between 1866 and 1877, and all were of the outmoded flush-decked type.

Glasgow & South Western Railway connections from Prince's Pier, Greenock were maintained by Captain Alexander Williamson's 'Turkish Fleet', so called by reason of the ships' names and their owner's star and crescent pennant. They sailed to Rothesay, Port Bannatyne and the Kyles of Bute. The most modern steamer in the 'Turkish Fleet' was the *Viceroy*, built in 1875, which was of raised quarter deck type, and propelled by a single diagonal engine, but the other three vessels dated from the sixties.

PRIVATE OWNERS

The largest fleet in 1888 was that of Captain Buchanan who, in the sixties, had been a partner of Captain Williamson, but the two had gone their separate ways, although continuing to share the same black funnel with broad white band which distinguished their steamers for many years. The Buchanan fleet comprised a variety of vessels, the oldest being the *Balmoral*, of 1839, and the newest the *Scotia*, built in 1880. Most had been purchased second-hand, and of these the most modern was the *Benmore*, acquired from Captain Robert Campbell in 1884 together with two older steamers. The *Benmore* was a straightforward example of the best practice of the mid-seventies, with nothing really outstanding about her, but she outlasted all of her contemporaries of the same general type.

The *Guinevere*, which entered service in 1869, was a two-funnelled steamer with raised quarter deck and oscillating machinery. Captain Buchanan bought her in 1884, employing her on the Glasgow–Rothesay route. His other two-funnelled steamer, the *Scotia*, however, spent practically all of her Clyde career on the Ardrossan–Arran route. She was an unusual steamer by Clyde standards, having a raised forecastle and a full-width deck saloon aft. With paddleboxes of striking design, and thin funnels, she was no beauty but maintained the Arran service reliably until the early nineties when she was completely out-classed by the new railway steamers. She had a double steeple engine, the last example of this very early pattern of marine engine to be used in a new Clyde steamer, and the only one with two cylinders. Buchanan's remaining three steamers were all acquired from other owners and dated back to the fifties and sixties, none being in any sense outstanding.

The Campbeltown and Glasgow Steam Packet Joint Stock Company Limited maintained a passenger and goods service between Campbeltown and Glasgow with three elegant single-screw steamers, the *Kintyre*, *Kinloch* and *Davaar*. The first-named, built by Robertson & Co in 1868, was widely regarded as the most attractive of the three; all had fiddle bows and their appearance suggested steam yachts of the time rather than commercial vessels. The *Kinloch*, built by A. & J. Inglis of Point-house in 1878 was similar, but larger, but the *Davaar*, built at Govan by the London & Glasgow Shipbuilding Company, Ltd in 1885, differed from her predecessors in having two funnels and a narrow deck saloon aft. All three steamers had two cylinder compound vertical engines by 1888, but the *Kintyre* had been single expansion until her conversion in 1882. The route followed by the Campbeltown Company's ships was from Glasgow to Greenock, Gourock (after 1888), Wemyss Bay, and thence via the south end of Bute to Lochranza, Pirnmill and Carradale to Campbeltown.

Messrs Hill & Co ran a service from Fairlie Pier to Millport and Kilchattan Bay in connection with the Glasgow & South Western Railway, using an old paddle steamer called the *Cumbrae*, which had appeared as the *Victory* in 1863. Of flush-decked design, with a steeple engine, she was obsolete by the late eighties and survived only until new railway vessels appeared during the following decade.

THE CAMPBELLS OF KILMUN

One of the best known and most popular of all the private steamer owners on the Clyde was Captain Robert Campbell—'Bob Campbell of the Kilmun boats'—who for many years maintained a service between the Broomielaw and the Holy Loch piers. There had been financial troubles in 1884 when his whole fleet was sold to Captain Buchanan but the support of his many friends enabled him to buy the *Meg Merrilies* almost at once. With this vessel he repulsed the invasion of his territory and retrieved his fortunes to such good effect that new steamers were added in 1885 and 1886, and with the able assistance of his sons Peter and Alexander the family business was soundly rebuilt. The new vessel of 1885 was the *Waverley*, built by McIntyre of Paisley and engined by Hutson & Corbett. She resembled the *Meg Merrilies* in having a full deck saloon aft, but none forward. She was found to be unsuitable for the daily Glasgow–Kilmun service and therefore appeared frequently on excursion duties. Her Clyde career ended quickly for, in 1887, she went on charter to the Bristol Channel and traded so successfully that the Campbells transferred their business there when the Caledonian Railway started to operate Clyde services in 1889. To replace the *Waverley*, the *Madge Wildfire* appeared on the Kilmun route in 1886 and she, with the *Meg Merrilies*, maintained the Holy Loch services in 1888.

Of all Clyde steamboat proprietors, Bob Campbell was the most vulnerable to railway opposition, for the new Caledonian pier at Gourock lay just across the firth from Kilmun and the other

Holy Loch piers. Nevertheless he was not required to face up to the problems of the new age, for after a long illness he died at his home in Glasgow on 10 April 1888, at the early age of 58. Many tributes were paid to his memory, and to his kindness and generosity. The funeral took place on 13 April, and was on a scale never repeated on the Clyde. Following a religious service at his home at 2, Parkgrove Terrace West, Bob Campbell's remains . . . 'were afterwards conveyed to the Broomielaw, followed by a large number of well-known citizens representative of the shipping interest and others. The coffin having been put on board the steamer *Madge Wildfire*, that vessel afterwards proceeded to Kilmun, calling *en route* at Greenock, Kilcreggan, and other coast piers, the company being thus largely augmented. At Kilmun religious services were conducted in the saloon of the steamer . . . and at the grave. . . . The coffin, of polished oak, which was wrapped in a Union Jack, was carried to the grave by eight members of the steamboat's crew, followed by the mourners, who numbered several hundreds. The bell of Kilmun Church was tolled, and the residents testified their respect for deceased in many ways. . . . The flags of many of the steamers in the harbour were displayed half-mast high, and the same mark of respect was likewise shown by most of the vessels at Greenock. . . .'

Bob Campbell's death symbolised the passing of an era in Clyde steamer history. He and his kind had served Glasgow and the Clyde well over a long period, but they were leaving the scene; within the following five years Captain Buchanan and Captain Alexander Campbell of the Wemyss Bay fleet were dead too, and Captain Alexander Williamson had retired. The day of the private owner as understood in the sixties and seventies was finished and for the few who remained in business the nineties brought competition on a scale never experienced previously. Through the summer of 1888 the new Caledonian pier and station at Gourock took shape, heralding a new pattern of Clyde passenger services. The brothers Campbell were intelligent businessmen as well as competent seamen. Their future was at Bristol, where the *Waverley* was a great success. Not for them a dour, rearguard struggle against the inevitable. Early in November, 1888 the papers reported that they had sold the *Meg Merrilies* and *Madge Wildfire*, with the goodwill attaching to the Kilmun trade, to the Caledonian Railway for the sum of £18,600.

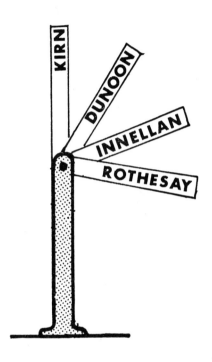

CHAPTER TWO

THE RISE OF THE CALEDONIAN
(1889-1894)

T HE directors of the Caledonian Railway did not at first fully realise the significance of their Gourock extension and attempted to make arrangements with private owners for the provision of steamer services in connection with the company's trains from Glasgow. After various rebuffs they set about acquiring ships of their own, only to meet with an embarrassing setback in Parliament, and were eventually forced to float an independent company to run the new steamboat services. Under the managership of the outstanding Clyde personality of the day, the Caledonian revolutionised steamer traffic on the firth, setting new standards of speed, comfort and efficiency with its fleet of modern vessels. Within three years the company was in a commanding position and appeared likely to establish a virtual monopoly on the river as traffic surged to its attractive new routes to the coast.

STEAMER PLANS

During 1888, with the prospect of the Gourock railway being opened in the early summer of the following year, the Caledonian Railway directors set about arranging a regular steamer service from their new railhead. At this stage it seems that they had no immediate intention of entering the field of steamboat ownership for their first step was to instruct the general manager, Mr (later Sir) James Thompson, to write to all Clyde steamer owners offering facilities and requesting that calls be made at Gourock. The overture had a mixed reception. Some owners, notably the Glas-

Page 35: Arran idyll: the *Duchess of Rothesay* disembarking passengers at Corrie Ferry.

Page 36: Excursion steamers of the eighties: *(above)* the *Ivanhoe* in her 'teetotal' days; *(below)* the *Victoria*, flagship of the Wemyss Bay Company.

gow & Inveraray Steamboat Company, did not even reply; David MacBrayne did so, offering to arrange calls by the Ardrishaig mail steamer; but none, in the view of the directors, gave the assurance of a service worthy of the new route.

It is possible that the Campbells of Kilmun influenced the board in its decision to abandon negotiations and commit the Caledonian Railway to steamer ownership on its own account, although Captain James Williamson of the *Ivanhoe* may well have encouraged its change of attitude too. Peter and Alexander Campbell's Kilmun service was vulnerable to competition from the Gourock railway route and the eclipse of their trade was inevitable. The success of the *Waverley* on the Bristol Channel offered prospects of a new business; the brothers saw their opportunity and grasped it. The steamers *Meg Merrilies* and *Madge Wildfire* were sold to the Caledonian as the nucleus of its new fleet, together with the goodwill of the Kilmun trade, and Peter Campbell agreed to remain with the new owners for a season or two before joining his brother at Bristol. The Caledonian board now acted swiftly and placed orders with Caird & Co, of Greenock, and John Reid & Co, of Port Glasgow for two new vessels in anticipation of Parliament's granting the necessary powers to operate a service. For this purpose the Caledonian Railway (Steam Vessels) Bill was introduced, empowering the company inter alia 'to provide and use steam and other vessels, and to raise additional capital', the amount estimated to be required being £66,000, with powers to borrow a further £22,000.

The Campbells were well thought of by the residents of the Holy Loch villages and their impending departure from the Clyde was marked by a pleasant ceremony on board the *Madge Wildfire* during her last sailing under the family's ownership on the afternoon of 31 December 1888. A large number of friends, under the chairmanship of Mr Cayzer, of the Clan line, presented a testimonial on behalf of passengers who had sailed with them for many years, and each brother received the gift of 'a clock with side ornaments'. Mr Cayzer spoke appreciatively of the Campbells and of their record of having never caused any loss of life in the conduct of their business, and voiced the general feeling of regret that they were giving up in favour of what was then regarded as a large public company with scant regard for local sentiment. He concluded by expressing the hope that the Caledonian Railway would offer a better service than had previously been given by public companies on the firth. In replying, Captain Peter Campbell regretted the severance of a family connection of thirty-five years' standing and spoke with gratitude of the many kindnesses experienced by his father during that time—no empty phrases these, for all present remembered the black times of the mid-eighties, when Bob Campbell's business was saved from ruin only by the timely and wholehearted support of his friends, an episode recalled by his son when he said that 'it had fallen to the lot of few men to receive such touching proofs of generosity and sympathy as had been shown to his father three years ago'.

The departure of the Campbells was widely deplored, for they represented the best amongst the private owners of the Victorian era. Had they remained as managers for the Caledonian Railway there can be little doubt that they would have created a service fully equal to the one that developed after 1888, but the steamers would have been rather different. The Clyde's loss, however, was the gain of Bristol, South Wales and the South of England where Peter and Alexander Campbell built up a fleet of magnificent paddle steamers in the years before the Great War. They always came back to Scotland for new tonnage, McKnight & Co of Ayr in particular being entrusted with the building of most of their beautiful ships, of which the *Westward Ho!* and *Britannia* were justly celebrated as the finest examples of their class in British waters.

The remarks of Mr Cayzer during the presentation to the Campbells reflected a widely held attitude towards the Scottish railway companies of the period. The travelling public and, more especially, railway and steamer enthusiasts of the present day tend to recall nostalgically the years when Scottish transport was controlled from north of the border, but our grandparents took a different view of concerns which they regarded as menaces to small private traders. It was felt by many people that railways in particular should be prohibited from owning steamers on the Clyde, where they were manifestly unwelcome to private proprietors, and operating outside their proper field in any case in the opinion of the public. Thus, at a meeting of the shareholders of the Caledonian Railway in Glasgow on 29 January 1889, held for the purpose of considering and approving the steamer Bill, there was trenchant criticism by the influential Mr McCaig, of Oban, of the principle of running in competition with existing owners. The latter shared his views fully and it was reported in *The Glasgow Herald* of 19 February that 'the Clyde Steamship Owners' Association resolved yesterday to petition against the bills promoted by the Caledonian . . . empowering them to provide and use steam vessels on the Clyde'.

A PARLIAMENTARY REBUFF

A parliamentary committee considered the Caledonian Railway (Steam Vessels) Bill in March 1889, when no less than fifty petitioners appeared against it. Objection was raised to this large number and eventually the committee admitted evidence from about a dozen, but the relatively wide powers sought by the company had alarmed many interests which had really no reason to fear the railway's intentions. Mr James Thompson stated the Caledonian case, which was that a great deal of money— £620,000—had been spent on the Gourock extension, exceeding the original estimates by fifty per cent, and naturally it was determined to ensure that such a massive outlay be recovered by making the best use of the new facilities. A lukewarm response from private owners to his circular letter of 1888 had induced his directors to reconsider their attitude. Apart from the first class tourist steamers and the vessels of the North British Steam Packet Company, the majority of Clyde steamers were out of date and the Caledonian Railway proposed to inaugurate a service with 'fast, clean and well-appointed steamers'. Private owners had had every opportunity to give a reliable service, but had made no offer to do so. The objection that fares would be increased by the railway was dismissed by Mr Thompson, who cited the competition of the North British and Glasgow & South Western companies as being sufficient to prevent this occurring. The remainder of the first day's hearing was taken up with supporting evidence by the Provosts of Dunoon and Rothesay, and other witnesses.

Evidence against the Bill was heard on the second day. The general feeling expressed by witness after witness was that existing private owners would be forced out of business by the intervention of the railway company. Malcolm T. Clark, secretary and manager of the Glasgow & Inveraray Steamboat Company, complained of a lack of precision in the circular sent to him by Mr Thompson, a point in which he was supported by Alexander Paterson, chief purser of the *Columba*, who represented David MacBrayne. Both witnesses said that they were prepared to arrange services to suit the railway, and Mr MacBrayne had already publicly announced his intention to arrange calls by the *Iona* and *Columba* at Gourock. Other evidence followed until the chairman intervened to say that the committee had made up its mind to grant at most only limited powers allowing a ferry service between Gourock, Kilcreggan, Cove, Kirn, Hunter's Quay, Dunoon, Innellan and Rothesay, and that further evidence should be given with this in mind. With prospects bleak for the promoters, the day's business finished.

The third day brought disaster. It started with additional evidence against the Bill and the case against the railway was powerfully argued. The Caledonian position was badly undermined by the damaging revelation that steamers had already been bought and ordered without parliamentary sanction and in spite of the efforts of Mr Pember, counsel for the company, the committee was clearly hostile. After summing up, the committee adjourned and, on resumption, the chairman stated that the preamble was unanimously found to have been not proven.

THE CALEDONIAN STEAM PACKET COMPANY, LTD

The Caledonian directors were now in a very embarrassing position. They had gone to some lengths to stress that their approach to Parliament was above reproach, but were now obliged to resort to a subterfuge. With only two months left before the opening of the Gourock route there was no time for manœuvre; the company had perforce to transfer its steamer interests to an associated company, by which transparent device it was possible to avoid having to seek parliamentary sanction. In fairness to the directors it seems that this course was morally distasteful as well as being administratively inconvenient, but they were obliged to bow to necessity. On 7 May 1889 the Caledonian Steam Packet Company, Limited was incorporated, having a share capital of £64,000 in 6,400 shares of £10, of which one each was taken up by seven railway directors. The first meeting of the new company was held in London on 21 May 1889, when the Marquis of Breadalbane was appointed chairman. The two Campbell steamers were re-registered in the name of the Steam Packet Company, having temporarily been registered in the names of three of the railway directors, since their purchase. Thus was launched almost by chance a venture that has existed until our own day, outlasting all of its contemporaries. Captain James Williamson, of the *Ivanhoe* syndicate, was appointed secretary and manager, with the registered office at 302, Buchanan Street, Glasgow. The original Caledonian steamboat committee had already chosen for the new fleet an official house flag, a yellow pennant with a red lion rampant, and the names *Caledonia* and *Galatea* had been decided upon for the two steamers under construction.

Of the two Campbell steamers acquired at the end of 1888, the *Meg Merrilies* was the larger, and had been renovated with a new boiler only a year before. She had also been given a surface condenser instead of her old jet condenser, and converted to the closed stokehold system with forced draught in an effort to cure her steaming troubles. In this modernised form she entered the

Caledonian fleet, the price placed on her for insurance purposes being £8,000. Her smaller sister, the *Madge Wildfire*, was valued at £6,000. The *Meg Merrilies* was probably the first ship to sail in the Caledonian colours. She was advertised to replace the *Madge Wildfire* on the Glasgow–Kilmun route on 1 April 1889 and it is reasonable to assume that she had been repainted in the new livery during the winter, whereas the *Madge*, constantly in service throughout the winter, would still be in Campbell's colours, save possibly for her funnel.

The Caledonian colours were very attractive and broke new ground on the Clyde. The hull was painted dark blue, virtually indistinguishable from black in bad light, and the underbody was sea green, with broad white boot-topping serving to lighten the whole appearance. Just below main rail level two gold lines were painted from stem to stern, being also carried round the sponsons. The main saloon was a delicate shade of pale pink, panelled in pale blue, this colour scheme being applied also to the sponson housings, while the wash boards round the promenade deck were white. The paddleboxes were painted white, with blue name boards, and there was much ornamentation and employment of gold paint. Paddle wheels were red lead. The deckhouses were varnished teak, and funnels Navy yellow, with no black top. The effect of this colour scheme, which owed something to contemporary naval practice, gave the impression of a private yacht, the crews being suitably uniformed to enhance the illusion. The smartness of Caledonian steamers soon became a byword; amongst older ships only the *Columba, Iona, Ivanhoe* and *Lord of the Isles* could rival them in the early days, but the example rapidly spread to other fleets, to the considerable benefit of the travelling public.

Construction of the Gourock railway was pushed ahead in the early months of 1889 and as early as 22 April—the Glasgow Spring Holiday—the *Meg Merrilies* was advertised to call at the new pier on her way from Glasgow to Kilmun. The old Campbell service was being continued by the new owners, the only alteration being the substitution of Gourock for Prince's Pier as a port of call. On 3 May it was reported that the contractors for the eastern section of the new railway had completed their work, and that it was expected that their colleagues on the western section would be able to finish their part of the line in time for the official opening on 1 June. On 4 May it was possible to run a special train over the entire route to allow the directors to see the progress of the work, and this was the first train to convey passengers between Greenock and Gourock. At last, with much publicity, the whole undertaking was declared open on 1 June. The first regular train to run over the new route started from Gourock at 5.25 am, conveying workmen to Greenock and Port Glasgow, and was driven by no less a person than Dugald Drummond, the company's celebrated locomotive superintendent, accompanied on the footplate by Mr John Cowan, a director both of the railway and the steam packet company. The first down train, also carrying workmen, was greeted with the waving of handkerchiefs, towels and tablecloths from windows of houses overlooking the line, and it was noted in a press report that passengers on the first train from Glasgow, which arrived at 7.20 am, 'met with a warm reception'.

The inaugural service from Glasgow Central to Gourock consisted of twenty-six down trains with an additional one on Saturdays, while a similar number in the up direction was increased to twenty-seven on Mondays only. The fastest run was done in forty minutes. In anticipation of the opening, Dugald Drummond had built at the Caledonian works at St Rollox a batch of three more of his small-wheeled 4-4-0 express locomotives, known variously as 'Greenock Bogies', or 'Coast Bogies', together with three new trains of ten carriages each, giving accommodation for 224 first class and 380 third class passengers. These were fitted with the Westinghouse air brake and 'all modern fittings'. In passing, it is of interest to record that the Caledonian, North British and Glasgow & South Western companies abolished second class travel to the Clyde coast after 1 March 1889 save only for the Wemyss Bay Railway section, over which the operating company, the Caledonian, did not as then exercise full control.

As the two new steamers being built for the Caledonian Steam Packet Company were not ready, the *Ivanhoe* was chartered from the Frith of Clyde Steam Packet Company to augment the two older steamers at the opening of the Gourock route. The *Ivanhoe* made the first steamer connection to Rothesay conveying passengers from the 8.30 am train from Glasgow. Traffic throughout the opening day, a Saturday, was heavy, particularly in the afternoon, and large numbers travelled to join the *Lord of the Isles* and *Columba* which also called at the new railhead.

NEW STEAMERS

Services from Gourock were augmented during the summer when the two new steamers entered traffic. The first, the *Caledonia*, had been launched by Captain Williamson's daughter on 6 May, and the new ship was completed in time to take up her sailings on 10 June, releasing the *Ivanhoe* for the Arran excursion, now modified to include calls at Gourock. The *Caledonia* was a revolutionary steamer for her time. The work of building her had been sub-contracted to John Reid & Co by the original contractors, Rankin & Blackmore, of Greenock, who built her boilers and machinery. It was with reference to the latter that the *Caledonia* represented a sharp break with traditional practice. The engine was of the compound type, having its two cylinders arranged in tandem fashion, driving a single crank, so that from the engine-room alleyway the onlooker could detect little, if any, difference from the older form of single diagonal machinery. The slide valves were worked by a slip eccentric, and a starting engine was provided to prevent the main one sticking on 'dead centre'. This machinery was a great advance, but the boilers were just as much of an innovation. There were two of them, of Navy, or locomotive type, working under forced draught in a closed stokehold instead of the older arrangement of a large haystack boiler working under natural draught. Naturally this installation cost much more to build than the traditional haystack and single engine, and in 1890 a correspondence developed in the columns of *The Glasgow*

Herald in which owners and builders—under *noms de plume*—assailed each other, arguing over the merits and disadvantages of the two systems. Additional construction costs in the *Caledonia* were offset by lower fuel costs and increased power, but her machinery and boilers were substantially heavier than the older type of installation, a factor which deterred other owners for many years. She was lavishly fitted and her general design was such that many naval architects seriously doubted her ability to maintain a satisfactory speed. They were discomfited when the *Caledonia* achieved 16¾ knots on trial, earning a premium for Rankin & Blackmore. While causing satisfaction by her performance, however, the new steamer inherited the main fault of the single engine—the unbalanced forces of her two pistons sliding back and forward caused the characteristic fore-and-aft surging motion associated with many of the older steamers.

This fault was not shared by her larger consort, the *Galatea*, which had been launched on 31 May and ran her trials on 3 July, attaining a speed of 17.36 knots. The *Galatea* was the flagship of the new fleet and in general appearance resembled the *Ivanhoe*, having two funnels placed fore and aft of the paddle-boxes. She was a considerable improvement on the older vessel in her machinery, being fitted with the first twin-crank compound engine used in a Clyde steamer. Steam was supplied by four Navy boilers working under forced draught. This arrangement caused the funnels to be placed very far part, giving the *Galatea* an ungainly aspect. She was intended to be an improvement on the *Ivanhoe* in every way, but was never as successful as her predecessor and her career in the Caledonian fleet was short. Nevertheless, she was the crack steamer of the day, and was placed initially on a daily cruise round the island of Bute, working in routine rail connections in the mornings and evenings.

The instant success of the Gourock route is evidenced by the following extract from *The Glasgow Herald* of 15 July 1889, describing the rush to the coast on Glasgow Fair Saturday:

Nothing could have surpassed the genial, exhilarating atmosphere of the morning. . . . At the Central Railway Station the passenger traffic throughout the day was heavier than on any previous Fair Saturday. Gourock alone was the destination of several thousand passengers, and last year so far as the Caledonian Railway is concerned, may be said to have had no existence. The *Cobra* train[1] for Gourock was sent off in two portions, each of 16 carriages, and every compartment was quite filled. So great was the rush of passengers for the *Cobra* train that bookings by the steamer had to be suspended twenty minutes before the advertised time of starting. Next came the *Columba* train, also despatched in two portions. The *Lord of the Isles* passengers filled one long train of carriages. The nine o'clock train to Gourock was run in two portions, as were also the ten o'clock, the 10.23, and the 11.35. Some idea may thus be formed of the enormous number of passengers who journeyed on Saturday over the Caledonian Gourock line. . . .

That the Fair Holiday traffic was no passing phenomenon was borne out by statistics quoted by Captain James Williamson at a social gathering of the employees of the Caledonian Steam Packet Company in Gourock early in 1890, when he stated that in the period from 1 June to 31 December 1889, nearly 700,000 passengers had been carried over the new Gourock railway. The new line was a success, and the Glasgow & South Western route via Prince's Pier now suffered the same fate which had befallen the original Caledonian route to the coast via Custom House Quay, Greenock, twenty years earlier on the opening of the Prince's Pier line. Traffic fell away dramatically and the decline confirmed the South Western directors in their resolve to press for powers to operate steamers on their own account and to make improvements in their company's coast services.

ARDROSSAN AND ARRAN

If the year 1889 had been one of triumph for the Caledonian Railway and its Steam Packet Company, 1890 brought still further advances on the lower reaches of the firth. During the late eighties, the Lanarkshire & Ayrshire Railway, a Caledonian

---

[1] In connection with the Belfast 'daylight' sailing, operated from Greenock and Gourock in 1889 instead of from Ardrossan.

protégé, had been projected to construct a new line from Lugton, on the Glasgow, Barrhead & Kilmarnock Joint Railway, and thence via Kilwinning, Stevenston and Saltcoats to Ardrossan, which it entered at the north side, and terminated at the harbour, where the Glasgow & South Western Railway or its earlier constituents had been ensconced since the 1840s. The main object of the promoters was to provide a route by which coal from the Lanarkshire pits could be taken to the Ayrshire coast for export to Ireland without carrying it over Glasgow & South Western metals, and it is doubtful if at any time they had considered the possibility of competing for the Arran traffic, then in the hands of Captain William Buchanan, in connection with the South Western company. Such potentialities seemed to lie more towards Ireland, and it was arranged with Messrs G. & J. Burns that they should inaugurate a 'daylight' service from Ardrossan to Belfast and back when the railway opened, as expected, in 1889 and the large paddle steamer *Cobra* was built specially for the new route. Unfortunately the railway contractors ran into difficulties in the reconstruction of Ardrossan Harbour and due to this and other circumstances the Lanarkshire & Ayrshire Railway was not fully opened until the summer of 1890. In the meantime, however, the Caledonian attitude to steamer ownership had altered completely following the events of 1889, and the imminent completion of the Ardrossan railway undoubtedly led to the decision to compete actively for the Arran traffic. Discussion of the project had evidently taken place in the early summer of 1889 for on 27 August the Steam Packet Company's minutes recorded that

A plan of the proposed steamer for the Ardrossan & Arran traffic was submitted and handed to the Marquis of Breadalbane, along with a copy of the specification.

By October, the directors' plans were sufficiently far advanced to allow consultation with the Caledonian Railway board and on 8 October the minutes recorded that

We issue specifications and invite tenders for the building of a steamer for the Ardrossan & Arran traffic.

Precisely one week later, a further minute recorded that

> The tenders for the building of the proposed steamer, for the Ardrossan and Arran service, were opened and read to the meeting viz :
>
> | | | |
> |---|---|---|
> | Messrs. | A. & J. Inglis | £37,000 |
> | „ | Fairfield Coy. | 31,750 |
> | „ | Scott & Coy. | 31,500 |
> | „ | Rankin & Blackmore | 27,500 |
> | „ | Denny Bros. | 24,400 |
>
> The meeting was adjourned . . . for further enquiry by Mr. Williamson *re* Messrs. Denny's tender.

And later the same day :

> Mr. Williamson reported the conversation he had had with the Messrs. Denny and Mr. Morton, with reference to compound and triple expansion engines, when it was agreed that Messrs. Denny were to submit an alternative tender . . . for what they considered would be an improvement on the proposed steamer, but their offer of £24,400 was accepted for the building of the steamer as specified.

The board met again on 22 October and its decision was minuted :

> The alternative offer from Messrs. Denny was submitted, *videlicet* the vessel to be 250′ × 30′ × 10′ moulded. To be propelled by a pair of fixed diagonal compound engines. Steam to be supplied by three boilers of the Navy type, having a working pressure of 110 lb per sq in, and all to be fired from one stokehole [*sic*]. The price to remain the same as their first tender, *videlicet* £24,400 for 17 knots. Premium to be at the rate of £3,000—above 17 knots, but under no circumstances is the cost of the vessel to exceed £27,400. Penalty to be at the rate of £3,000 under 17 knots, with power of rejection if not above 16 knots.
>
> It was agreed to accept the above offer.

Work on the new steamer was pushed ahead during the winter of 1889-90. On 21 January 1890 the minutes recorded that the first instalment of the price—£4,880—'will be due next week'. The second instalment was authorised to be paid on 1 April, and the third on 28 April, but in the meantime the new steamer was launched from Denny's no 9 slip at Dumbarton on the afternoon of 10 April. During the ensuing month her boilers and machinery were placed on board and by mid-May 1890 the vessel was ready for trials. She was named *Duchess of Hamilton* in honour of the lady of Arran, an appropriate choice in view of the service for which she had been designed and, as it transpired, the first in the series of 'royal' names which became a tradition in the Caledonian fleet.

While work was in progress it appears that a deliberate attempt was made to conceal the intention of the Caledonian company to enter the Arran trade, almost certainly to blind its Glasgow & South Western rival to the true state of affairs. Captain James Williamson was in private partnership with Robert Morton, practising as consulting marine engineers and naval architects under the name of Morton & Williamson; during the late eighties they had been jointly responsible for the design of two highly successful paddle steamers for service in Australian waters. When a rumour spread that the new Denny paddler was also intended for the Antipodes, nobody in Caledonian circles saw fit to correct it, and as late as 26 February 1890 *The Glasgow Herald* reported as follows:

> We understand the Caledonian Steam Packet Coy. have purchased from Messrs. Morton & Williamson, Glasgow, a paddle steamer now in course of construction in the shipbuilding yard of Messrs. William Denny & Bros., Dumbarton, and which steamer was intended for the Australian colonies. . . . The Caledonian Company intend to run the vessel from Ardrossan to Arran in connection with a service of fast trains from the city. . . .

It is not often appreciated that a Denny paddle steamer has always been a rarity on the Clyde—the next after the *Duchess of Hamilton* was the 1934 *Caledonia*—and consequently it was to be expected that the new vessel of 1890 would bear little resemblance to existing Clyde steamers. In fact, her resemblance to the well-known *Belle* steamers on the Thames—all Denny built— was more marked than any similarity to her Scottish sisters, save only the *Marchioness of Lorne* of 1891, which was designed to resemble her in many respects. The *Duchess of Hamilton* was the first Clyde steamer to have her promenade deck carried forward to the bow, although she was not plated in, to permit the mooring ropes to be worked from the main deck. This style of construction, although not peculiar to the Caledonian steamers, nevertheless came to be regarded as characteristic of the fleet. The after deck saloon was carried out to the full width of the hull, but the forward one was narrower to allow main deck alleyways round it. There were two small deckhouses on the promenade deck, one carrying the navigating bridge and the other combining the purser's office with a shelter on the stair leading down to the main deck. Although by modern standards accommodation on the promenade deck was Spartan, nevertheless the advance on most contemporary steamers was immense, particularly in regard to the generous deck space.

The *Duchess of Hamilton* ran her trials on 28 May 1890, her speed as a mean of two runs between the Cloch and Cumbrae lights being 18.1 knots, handsomely exceeding the guarantee of 17 knots stipulated in the building contract and earning the full premium of £3,000 for William Denny & Bros. On the following day she ran a special trip with a large party of invited guests, sailing from Gourock to Whiting Bay, Arran, via the Kyles of Bute and returning thence directly to Gourock. Lunch was served amid the state of euphoria usual on such occasions and many congratulatory speeches were made by the obviously well pleased builders and owners. An opportunity was found of lying in wait for the *Columba* off Kirn on the homeward run, and the new *Duchess* succeeded in beating her in a race to Gourock, although it seems that the MacBrayne flagship must have been off form on this occasion.

The Glasgow Herald enthused over the *Duchess of Hamilton*'s fittings:

The first class saloon is aft. . . . The woodwork . . . is of walnut with mahogany pilasters, enriched with hand painted gilt panels. The ceiling is very pretty, the colours being cream and gold. Spring-stuffed settees, affording a most luxurious seat, are placed athwartships in the saloon. There are also a writing and reading table, while the heating is effected by means of a very handsome stove. The floor is laid with carpets and carpet runners; the curtains and spring blinds, in terra cotta and tan, harmonise well with the general decorations. At the aft end a sliding door, having a stained glass panel with a portrait of the Duchess of Hamilton, admits to the quarter deck. The dining saloon is underneath and has seats for ninety persons. The cushions are in old gold frieze velvet, and the curtains on the ports are in blue and tan silk damask. The table covers are in silk tapestry, woven specially to the size of the table, and having gold borders all round.

The new Caledonian route to Arran via Ardrossan was opened on 30 May 1890. The importance attached to the new service was evidenced by the fact that a train of eight new bogie carriages was built specially for it to the designs of Dugald Drummond and given the name of *The Arran Express*. The engine allocated to its working was Drummond's 4-4-0 No 124, a celebrated locomotive of the period. She had been exhibited when new in 1886 at the Edinburgh International Exhibition of Science, Industry and Art by her builders, Dübs & Co, of Glasgow, and in 1888 had been chosen for royal train duties on the occasion of the opening of the Glasgow International Exhibition by the Prince and Princess of Wales. Now, in 1890, she was sent to the Ardrossan route and given the name *Eglinton*, in honour of the Earl of Eglinton, on whose estates the town of Ardrossan was situated, and whose family name—Montgomerie—was used for the Caledonian pier at the harbour. Naming of Caledonian locomotives was done only on rare occasions and was a further indication of the value placed on the new service by the board.

The inaugural train left Glasgow Central station at 4.45 pm, called briefly at Eglinton Street (Glasgow) for ticket collection, and then ran non-stop to Ardrossan. The trip was marred by a late arrival at Montgomerie Pier, and more time was lost in transferring luggage to the steamer so that the Glasgow–Brodick run was made in 108 minutes instead of the scheduled 90. An immediate instruction was issued that in future luggage for the *Duchess of Hamilton* should be despatched by an earlier train to avoid delays, and it was not long before the new schedules were kept regularly and even improved upon, the Glasgow–Brodick journey being covered on occasions in as little as eighty-seven minutes.

The *Duchess of Hamilton* came into immediate competition with Captain William Buchanan's steamer *Scotia*, then sailing from Ardrossan in connection with the Glasgow & South Western Railway and despite the Caledonian vessel's overwhelming advantages in comfort and speed the contest was not without its moments of drama. On the afternoon of 29 August 1890 the *Scotia* left Brodick ten minutes ahead of the *Duchess* which, however, gradually overhauled her until, at the approaches to Ardrossan Harbour, the larger steamer drew slightly ahead. At the breakwater, where both vessels passed through a narrow entrance, the *Scotia* rammed her opponent's paddlebox, forcing her out of course and nearly causing her to collide with a barge lying off Montgomerie Pier. At a subsequent court case in 1891, Captain Robert Morrison of the *Duchess of Hamilton* was blamed for the collision and damages of £32 2s were awarded to Captain Buchanan.

WEMYSS BAY IMPROVEMENTS

While the *Duchess of Hamilton* established Caledonian supremacy in the rival stronghold, events of importance were also taking place further north, at Wemyss Bay. For some time the relationship between the Caledonian Railway, which operated the independent Wemyss Bay Railway, and Captain Alexander Campbell, owner of the Wemyss Bay steamers sailing in connection with the trains, had been uneasy. The railway company was accused of unfairness in allocating a proportion of the through rail and steamer bookings to Captain Campbell while he, in turn, was said to maintain the regular services with inferior steamers while using the modern *Victoria*, a handsome two-funnelled saloon vessel, for cruising charters, which were more lucrative. No doubt there was a measure of truth on both sides, but apart from the steamer

Page 53 : The Inveraray route: the second *Lord of the Isles*, in original condition, arriving at Rothesay.

Page 54: Steamers of 1892: *(above)* the *Neptune* on cruising duties at Ayr; *(below)* Buchanan's *Isle of Arran* approaching Rothesay.

facilities the Caledonian Railway had long regarded the Wemyss Bay Railway as a thorn in the flesh. The principal shareholder was a Mr Lamont, and on several occasions litigation had taken place to settle differences between him and the Caledonian Railway. The ultimate sufferers were the travelling public, not the least of whose inconveniences was the fact that all Wemyss Bay trains used Glasgow Bridge Street station, situated on the south side of the Clyde, well away from the main business centre of the city. The Wemyss Bay route was not being exploited to its full advantage and it was generally expected that changes for the better would be made when, in August 1889, it became known that the Caledonian Railway had acquired the Wemyss Bay company on the basis of a straightforward exchange of ordinary stock. Improvements thereafter came speedily; on and after 1 May 1890 the Wemyss Bay trains ran to and from the Central station instead of Bridge Street, a change which had been sought for years. Almost simultaneously there came a crisis in the affairs of the steamer company. Evidently sensing that the successful flotation of the Caledonian Steam Packet Company meant that his days were numbered, Captain Alexander Campbell unexpectedly gave notice in April 1890 of his intention to withdraw the steamers at the end of the month. His contract was due to expire in September of that year but doubtless Captain Campbell hoped to force better terms while the Caledonian Railway was at a disadvantage. The reason which he gave for this drastic step was that the new Gourock route had drawn away so much of the coast traffic that the proportion left to the Wemyss Bay fleet was insufficient to allow the services to be conducted efficiently as well as economically.

Unfortunately for Captain Campbell, James Williamson was equal to the situation. Notwithstanding the ultimatum, the May timetable provided for a full connection of steamers from Wemyss Bay to Rothesay, Millport and Kilchattan Bay. The company was in fact much more favourably placed to take over the service than might have been supposed, for in addition to the *Duchess of Hamilton* two smaller steamers were under construction and one had already been launched on 15th April. The immediate problem was therefore to keep things going between the end of April and the arrival of the new ships, which in practice was done without difficulty by switching one of the Gourock steamers to Wemyss Bay and re-arranging services temporarily, augmenting as necessary by chartering.

The two new steamers were contracted for by Rankin & Blackmore, of Greenock, who supplied the boilers and machinery, but sub-contracted the building of the hulls to John Reid & Co, of Port Glasgow. They were sister ships, based on the design of the *Caledonia*, but incorporating modifications and improvements resulting from experience with the earlier steamer. Boilers and engines were of the same type, but steam pressure was slightly higher than in the *Caledonia*. From the observer's point of view, the most radical break with tradition was the placing of the navigating bridge forward of the funnel, a feature which also distinguished the *Duchess of Hamilton* and thenceforward became standard in ships built for the company. The older practice of designing paddle steamers with bridges between the paddle-boxes was not derived from caprice or ingrained conservatism on the part of owners. The small steamers of the Victorian period responded well to skilful handling at piers and it was found that a position on the paddlebox was superior to any other for man-œuvring work. The view ahead past the thin funnels of those days was perfectly adequate for all navigational purposes and consequently there arose no demand for a change. Nevertheless, a series of disquieting incidents occurred in 1889, involving steamers of the Caledonian Steam Packet Company and its associates, drawing attention to the dangers of inadequate lookout in certain circumstances. On 20 July the *Caledonia* ran down a small boat containing four holidaymakers while leaving Craigmore Pier, Rothesay, and while fortunately there were no fatalities there was a public outcry about the danger to small boats which inadvertently crossed the paths of steamers. The Wemyss Bay steamer *Adela* was involved in a more serious affair in August involving the death of a Glasgow doctor, when the boat which he was sailing with a friend was run down near Toward, and a newspaper correspondence ensued in which there was a strong public demand for the repositioning of bridges forward of funnels so as to secure a better lookout. Scarcely had the furore subsided when the *Madge Wildfire*, crossing from Kilcreggan to Gourock on the evening of 28 September, collided in the darkness with the small private steam yacht *Osprey*, sinking her and causing the deaths by drowning of three out of her crew of four. The yacht was carrying no lights and it was hardly surprising that Captain Peter Campbell on the *Madge Wildfire* failed to see her crossing his bow. Nevertheless, it was the worst in a succession of similar accidents and it is inconceivable that the new steamers of 1890 were built with bridges forward merely by coincidence; almost certainly Captain James Williamson was influenced in this matter by the tragedies of 1889.

The first of the 1890 sisters to be launched was named *Marchioness of Breadalbane*, a graceful compliment to the chairman's lady, while the second, launched by Miss Maud Williamson on 6 May, received the name *Marchioness of Bute*. The *Marchioness of Breadalbane* ran her trials on 21 May, achieving a speed of 17 knots, while her consort repeated this performance on the Gareloch mile on 6 June, both vessels earning a premium of £150 for Rankin & Blackmore by virtue of having exceeded the contract speed of 16¾ knots.

The summer of 1890 saw the Caledonian firmly entrenched at three Clyde railheads—Gourock, Wemyss Bay and Ardrossan—and bidding fair to establish a monopoly on the south banks of the firth. The Glasgow & South Western Railway was taken at a disadvantage—Prince's Pier by-passed by the new and attractive Gourock route, its Ardrossan stronghold invaded, and the Millport and Kilchattan Bay service at the mercy of a revitalised thrust from Wemyss Bay. The only part of the coast where it still held an advantage was in respect of the shorter distance to Millport from the railhead at Fairlie. It was a grave situation, which the company could not immediately remedy, for in 1890 it had to fight a parliamentary contest in which first the North

British and then the Caledonian attempted to absorb it. The North British scheme was supported by the South Western directors, but the massive opposition of shareholders, organised and led by the shipowner Sir John Burns, ultimately defeated the attempt. Nevertheless the company was prevented for several months from trying to recover its Clyde traffic and it was not until 1891 that the first steps were taken with a view to restoring the position.

The extent to which the public flocked to the Caledonian was shown by *The Glasgow Herald*'s description of traffic from Glasgow on Fair Saturday, 1890 :

> The fine weather by which the Fair holidays were ushered in happily continues . . . it was on the coast traffic that the increase at the Central was most marked. Throughout the forenoon yesterday the ordinary trains to Greenock and Gourock had each to be run in two portions; and in the afternoon no fewer than fourteen specials were required to accommodate the traffic. Two of the afternoon trains for Wemyss Bay were duplicated, one had to be run in three and one in four sections, and two special trains were despatched in the evening. . . . With the increased facilities presented by the railway companies the traffic from the Broomielaw was lighter than in some former years. The morning steamers were well filled, but as the day advanced the numbers decreased, and the late boats had comparatively few passengers.

Not content with their successes, the Caledonian Steam Packet Company next considered the matter of the Clyde winter services. Summer traffic demanded steamers of the size and carrying capacity of the *Duchess of Hamilton*, but these were highly unprofitable to operate during the winter months; most were withdrawn annually in the autumn and laid up until the following year. The decision of the Caledonian not to use the *Duchess of Hamilton* on the Arran service during the winter of 1890-91 caused inconvenience and drew sharp comment from newspaper correspondents, some of whom evidently saw all the evils of railway monopoly eventuating sooner than they had anticipated. There was some bitterness against a company which skimmed the cream of the traffic in the high season and abandoned the unprofitable part of the year's working to Captain Buchanan, who by his faithful support of the Arran public throughout the year should have been entitled to better fortune in the summer. Unknown to these writers, however, the Arran winter service had been considered by the Caledonian board for several months. A minute of 1 July 1890 read as follows:

> Tenders from Messrs. Denny and Rankin & Blackmore were submitted for the building of a Twin Screw Steamer for service between Ardrossan and Arran in winter. Agreed to delay in the meantime.

The main interest of this minute lies in the board's idea that a twin screw steamer might be superior to a paddle steamer for the work envisaged. Had this intention materialised the resulting vessel would have been the first of the type to sail in Clyde passenger service, but as events transpired, no twin screw steamer was ever built for the Clyde. The idea of a screw steamer was again discussed at a board meeting on 19 August 1890 but an alternative suggestion for a paddle steamer was put forward and eventually approved. A minute of 21 October 1890 recorded that

> Tenders for the building of a Paddle Steamer were submitted from Messrs. Caird & Co. £21,250, Denny & Bros. £21,000, Scott & Co. £18,350, Fleming & Ferguson, £19,800, Fairfield Shipbuilding Coy. £18,540 (Tandem £17,450) Rankin & Blackmore £15,850. It was agreed to accept the Tender of Messrs. Rankin & Blackmore (along with Messrs. Russell & Co. for the building of the hull) for the sum of £15,850.

The cost of the new vessel was financed by a loan of £16,000 from the Caledonian Railway to the Steam Packet Company.

The name of the steamer was discussed by the board on 3 February 1891 and the Marquis of Breadalbane was authorised to seek the permission of H.R.H. Princess Louise to use her title *Marchioness of Lorne*, and this was duly granted. This recalls the years when the Duke of Argyll was prominent in national politics and there had been widespread satisfaction when his heir, the Marquis of Lorne, married Queen Victoria's daughter. 'Wassna the Queen the proud woman,' one of the Inveraray worthies is reputed to have said, 'when her daughter merrit the son of the Duke of Argyll!' The choice of name for the new Caledonian steamer was therefore a popular one and continued the tradition of Caledonian nomenclature established with the Duchess and two Marchionesses of 1890.

NEW ENGINES AND BOILERS

The *Marchioness of Lorne* was an unusual vessel, even in the Caledonian fleet, and marked another stage in the quest for mechanical and thermal efficiency. Instead of a compound engine, she was driven by a triple expansion set of machinery, the first to be used in a Clyde steamer. Whereas in a compound engine of the normal two-cylinder type steam exhausted from the high pressure cylinder is used again at lower pressure in a larger, low pressure cylinder, in a triple engine the process is carried a stage further and the steam, at still lower pressure, passes into a third cylinder to do more work before finally exhausting to the condenser. The potential economies of such machinery had long attracted James Williamson and he and Robert Morton had jointly been concerned in its application to the steamers built for Australian service in the late eighties. The usual form of triple engine in a paddle steamer involves the use of three cylinders and cranks—the standard form of later years—but the type of machinery employed in the *Marchioness of Lorne* was quite different and was in effect a combination of the compound engine of the *Duchess of Hamilton* and the tandem compounds of the *Caledonia* and the two Marchionesses. The intermediate and low pressure cylinders were placed side by side as in the conventional compound arrangement and drove two cranks, but working in tandem with each was a high pressure cylinder. This ingenious scheme resulted in a considerable saving of space and allowed one crank and connecting rod to be dispensed with as compared with an ordinary three cylinder triple engine. From the engine room alleyway it was impossible to detect any difference between the *Marchioness of Lorne*'s machinery and that of any ordinary two cylinder compound vessel. Steam was supplied by two Navy boilers, working as usual under forced draught.

Outwardly the *Marchioness of Lorne* was built to resemble the *Duchess of Hamilton* rather than the earlier Marchionesses and the *Caledonia*, although she was virtually the same length as the older trio, but with greater beam. Unlike them, her promenade deck was carried forward to the bow in the same style as the *Duchess of Hamilton* and her resemblance to her larger consort was emphasised by her paddleboxes being of Denny pattern despite the fact that her builders were Russell & Co of Port Glasgow. Another feature shared only by these two vessels in the Caledonian fleet was white deckhouses, panelled in pink and blue, all other steamers having varnished teak deckhouses. The *Marchioness of Lorne*, while always a useful and popular steamer, was no beauty. She shared most of the aesthetic faults of the *Duchess of Hamilton*—lack of sheer, too upright a funnel, and badly placed navigating bridge—and added some of her own, of which the undue length of the promenade deck forward of the saloon was a notably ugly characteristic, although this feature was much improved in later years when her saloon was lengthened by several feet.

The *Marchioness of Lorne* was on trial on 22 June 1891 but her speed disappointed. She was reported as having reached only sixteen knots, which fell short of the guaranteed speed by a quarter of a knot, and Rankin & Blackmore as principal contractors suffered a penalty of £100. Despite this failure, the new steamer was not otherwise unsatisfactory. The Arran winter service, her *raison d'etre*, did not call so much for speed as seaworthiness, and she had been built strongly in view of the wild weather so frequently encountered on the crossing from Ardrossan to Brodick. In summer her duties were those of spare boat and excursion steamer, although for a time she became identified with the Wemyss Bay and Millport service. Her accommodation was well suited to cruising, for she had roomy decks and saloons, while

EVENING CRUISE

And the Most Magnificent
DISPLAY OF
FIREWORKS

EVER EXHIBITED IN SCOTLAND.

Also, the Illumination of the Woods of Kelly

AT

WEMYSS BAY,

By Messrs BROCK, of London,

On **FRIDAY EVENING, 22D AUGUST, 1890**

(Weather favourable).

Messrs C. T. BROCK & Co.'s Leading Successes of the Year.
NEW SPECTACULAR FIREWORK SCENE,
"MAN THE LIFEBOAT"
(The Great Crystal Palace Sensation);
THE FIREWORK MARIONETTE SKELETON
(The Latest Novelty in Firework Mechanical Pieces);
THE WHISTLING FIREWORKS
(Brock's Latest New Patent);
Moving Devices, Fixed Pieces, Huge Shells, Jewelled Clouds,
Flights of Rockets, Shooting Stars, Dragon Flies, Meteors,
Magical Illuminations, and a host of Brilliant Pyrotechnic
Devices.

SPECIAL TRAIN from GLASGOW (CENTRAL) to
WEMYSS BAY at 8.11 P.M.; Returning from Wemyss Bay
at 10 P.M.
RETURN FARES—1st Class, 3s; 2d Class, 2s 5d; 3d Class, 2s.
Trains from GLASGOW (CENTRAL), to GOUROCK,
6.30 and 7 P.M.; Returning from Gourock at 10.30 P.M.
RETURN FARES (including Steamers), 2s 4d and 3s.

STEAMBOAT ARRANGEMENTS.

"DUCHESS OF HAMILTON"
(Lit up throughout by Electricity),
From PRINCES PIER, GREENOCK, at 7.30 P.M. ;
GOUROCK, 7.50 P.M.
"IVANHOE,"
From HELENSBURGH at 7.30 P.M.; GOUROCK, 7.50 P.M.
(Season Tickets are not available on this run).
"MARCHIONESS OF BUTE,"
From KIRN at 8 P.M.; DUNOON, 8.10 P.M.;
INNELLAN, 8.25 P.M.
"MADGE WILDFIRE,"
From KILCREGGAN at 7.55 P.M.; COVE, 8.5 P.M.;
BLAIRMORE, 8.13 P.M.; STRONE, 8.22 P.M.;
ARDENADAM, 8.30 P.M.; KILMUN, 8.37 P.M.;
HUNTER'S QUAY, 8.45 P.M.
"MEG MERRILIES,"
From ROTHESAY at 8.30 P.M.; CRAIGMORE, 8.35 P.M.
"CALEDONIA,"
From LARGS at 7.30 P.M.; MILLPORT, 7.50 P.M.
"GALATEA,"
From WHITING BAY at 7 P.M.; KING'S CROSS, 7.10
P.M.; LAMLASH, 7.25 P.M.; BRODICK, 7.45 P.M.;
CORRIE, 8.5 P.M.
RETURN FARE—One Shilling.
Season Tickets are not available.
For Programme of Fireworks, see Handbills.
Caledonian Steam Packet Company, Limited,
302 Buchanan Street, Glasgow.

Fireworks at Wemyss Bay—a Caledonian extravaganza of 1890

her furnishings drew the particular attention of *The Glasgow Herald*'s reporter :

> The drawing saloon is a large airy compartment extending the entire width of the vessel, and it is magnificently furnished in highly polished mahogany. The sumptuous appearance of the whole is accentuated by the pilaster panels, which are most artistically decorated with varied designs in hand-painted gilt. The general effect is further heightened by the ceiling, which is painted cream colour, beautifully relieved by white and gold, forming a strong yet harmonious contrast to the rich red-tinted Brussels carpet and runners of quaint Indian design. . . . The woodwork (of the first class dining) saloon is also of polished mahogany, and the ceiling is prettily painted in pale green and gold. . . .

The new Marchioness was ready for service at the beginning of the summer season of 1891, a year in which Captain James Williamson and the Caledonian directors must have considered that it was only a matter of time before the bulk of Clyde traffic was in their hands. Glasgow Fair traffic again broke all records and it was recorded that the pressure on coast services was intense, most of the ordinary trains from Central station on Fair Friday requiring to be duplicated while 'between four and six o'clock those for Gourock and Wemyss Bay had to be run in three, four, or five sections.' We have other glimpses of the Caledonian fleet in that summer, one of the highlights being the chartering of the *Galatea* by the Royal Clyde Yacht Club to act as club steamer during the Clyde Fortnight, in the early part of July. The new *Marchioness of Lorne* acted in quite a different capacity when she was placed free of charge at the disposal of Millport Visitors' Club also in July, when a party of five hundred enjoyed an evening cruise round Bute, the proceeds going to the funds of the club. We learn that 'when crossing Rothesay Bay on the way home a concert was started in the saloon . . . several young ladies gave their services, which were greatly appreciated. A most enjoyable evening was spent, the steamer arriving back at Millport about half past ten.'

The *Duchess of Hamilton* was involved in a more dramatic

occasion on 5 August 1891. An excursion had been arranged from Ayr to Inveraray, and there was a large complement of passengers, estimated as at least one thousand people. It was the *Duchess*'s first trip up Loch Fyne, and as she approached Inveraray those on board were appalled to discover that the steamer was running at full speed on to a sandbank known as the Otter Spit, where she stuck fast and could not be got off, as the tide was ebbing. *The Glasgow Herald* reported that 'it is understood that the accident was due to a slight misunderstanding between the captain and the pilot'! Here was a fine situation—the Caledonian's crack steamer aground in broad daylight in full view of the amazed citizens of Inveraray. Fortunately there was no danger and the alarmed passengers were ferried ashore by relays of small craft, including the Inveraray and St Catherines paddle ferry *Fairy*. The weather was fine, and the stranded travellers lit bonfires on the shore as the evening advanced. Inveraray post office was besieged and over 800 telegrams were despatched to Ayrshire containing reassuring messages to anxious relatives. In the meantime the *Duchess of Hamilton* floated off on the flood tide after ten o'clock and as no damage had been caused by her stranding, the passengers were re-embarked and she set off for home. Ardrossan was reached at 3.15 am and Ayr at 5 o'clock, where one of the 'honest men' remarked on landing that 'it was an awfu' guid twa bob's worth' but he would prefer less the next time! Like Queen Victoria on another famous occasion, the Caledonian directors were not amused, and a minute of 18 August reported

> the accident to the 'Duchess of Hamilton' at Inveraray . . . and it was resolved to censure Captain Morrison for his error in judgment, but under the circumstances it was agreed to allow him to remain until the end of the season, when the question would be further discussed.

It was clear that the captain had come close to losing his post, but on 28 October the board met again and resolved 'to retain his services in respect of his general ability and otherwise satisfactory conduct'. Although a strict disciplinarian, Robert Morrison was one of the best captains in the company's service, and the directors evidently recognised his qualities and were prepared to overlook the unfortunate, but spectacular, affair at Inveraray.

With the introduction of new steamers the shortcomings of the older vessels in the Caledonian fleet became more apparent. Both of the Campbell boats, the *Madge Wildfire* in particular, required modernisation to bring them up to the new standards. The *Madge* had sailed winter and summer continuously since 1886 and in July 1890 it was reported that her boiler needed a major overhaul. Captain James Williamson recommended reboilering and compounding of her machinery and the offer of Rankin & Blackmore to do the work at a cost not exceeding £2,500 was accepted in September. This was the first instance of a Clyde steamer being converted from simple to compound expansion and the results were highly successful. The original single engine had a high pressure cylinder added, to produce a tandem compound of the type used in the *Caledonia*, and two Navy boilers, working at a pressure of 100 lb per sq in in a closed stokehold replaced the haystack of the original design. The opportunity was taken of adding a short fore saloon. Thus renovated the *Madge Wildfire* was to all intents and purposes a slightly smaller version of the *Caledonia*. She went on trial on 12 March 1891 when it was found that her original speed could be maintained indefinitely at a much lower fuel consumption than before. The *Meg Merrilies* was the next to be dealt with; late in 1892 Barclay, Curle & Co's offer to reboiler her at a cost of £1,490 was accepted. It is not entirely clear what pattern of boiler replaced the existing haystack but the machinery was not converted to compound during the reconstruction and it seems most probable that the new boiler was of the older, but well-tried, design.

The reboilering of the *Meg Merrilies* brought to a close the first phase of the Caledonian Steam Packet Company. Within the space of four seasons the Caledonian Railway, through its subsidiary, had virtually transformed the Clyde steamer scene and come not far short of establishing a monopoly on the river. Under Captain James Williamson there had been created a fleet of modern vessels of a uniformly higher standard than any on the firth save only for the tourist steamers *Columba* and *Lord of the Isles*. Machinery and boiler design had been advanced more than in the twenty years before 1889. The Caledonian company now owned the largest individual fleet on the Clyde, providing undoubtedly the best service. Under a vigorous, thrusting management the future seemed assured and prospects bright. Nevertheless competitors were not overawed and the pace set by the Caledonian acted as a spur to the Glasgow & South Western directorate, who had viewed with alarm the inroads made by the newcomer into their traditional fields. Unabashed, they set themselves the task of remedying the situation, and it is to their affairs that we must now turn our attention.

CHAPTER THREE

SOUTH WESTERN RESURGENCE
(1890-1894)

THE Glasgow & South Western Railway was badly affected
by the successful Caledonian drive to the Clyde coast but
for various reasons it was unable to take immediate steps
to remedy its position. When eventually it sought parliamentary
powers to run steamers they were granted with certain restrictions.
Within its limits the company fought to recover lost traffic,
modernising train services and Clyde railheads, and providing,
regardless of expense, a fleet of superb modern paddle steamers
which competed always vigorously and sometimes recklessly with
the Caledonian vessels. By the mid-nineties the company's efforts
had been largely successful in retrieving a situation which had
seemed lost only a few years before.

PARLIAMENTARY BATTLES

The directors of the Glasgow & South Western Railway were
under no illusions as to the effect of the Caledonian manœuvres
of 1889-90, for the effect upon their traffic was immediate and
devastating. It was calculated that bookings from Glasgow and
Paisley to the Clyde coast had fallen from 291,300 passengers in
1888 to 158,766 in 1889 and 101,437 in 1890 and revenue from
£11,211 in 1888 to £6,108 in 1889 and £3,850 in 1890. Traffic
at Prince's Pier, Greenock was particularly badly affected by the
opening of the Caledonian's Gourock extension. Passengers
exchanged with the *Lord of the Isles, Chancellor, Edinburgh
Castle* and *Windsor Castle* fell from 34,891 in 1888 to 15,230 in
1890 while the traffic exchanged with the *Ivanhoe* slumped from

67

12,441 to only 1,576 in the same period. At Ardrossan the superiority of the *Duchess of Hamilton* over the Buchanan steamer *Scotia* was reflected in a drop of no less than seventy-five per cent in the Glasgow & South Western Railway's Arran traffic. Captain Buchanan was in no position to continue such an unequal contest and informed the railway company of his intention of withdrawing the *Scotia*. The South Western board had no alternative but to offer to guarantee him against loss and in exchange for maintaining the Arran winter service of 1890-91 he was paid £100 per month in addition to the normal consideration.

Captain Alexander Williamson, owner of the 'Turkish Fleet', sailing in connection with the South Western trains at Greenock, found himself similarly embarrassed by the superiority of the Caledonian steamers under his son's management at Gourock. He too warned the railway company that he could no longer continue the service and again negotiations had to be entered into to safeguard it in the immediate future until the South Western took steps to compete more effectively. Captain Williamson was therefore paid the sum of £1,500 for the 1890 season over and above the steamboat proportion of through bookings to the coast.

The directors were well aware of the parliamentary attitude towards ownership of steamboats by railway companies. Only two years had elapsed since the Caledonian attempt had been frustrated and there was little reason to expect that a South Western Bill would be any more successful. However, the Glasgow & South Western Railway (Steam Vessels) Bill was drawn up with care and in the hope of minimising hostility amongst the powerful steamboat interests. Clause 4 of the Bill expressly provided that 'powers shall not extend or apply to traffic to or from Inveraray, Ardrishaig, Tarbert, or Campbeltown'—an obvious concession to David MacBrayne, the Inveraray company and the Campbeltown & Glasgow owners, but it failed in its object for they joined the other members of the Clyde Steamship Owners' Association in petitioning Parliament against the Bill.

The powers sought to be obtained under the bill would enable the company directly and injuriously to compete with many of your petitioners, who supply ample accommodation for the wants of the public. . . . It is most inexpedient and objectionable to grant large powers of owning and using steamboats to railway companies. The result of granting such powers has been to enable them to use their great pecuniary resources, as well as the control of the traffic which they possess in the working of their own lines, to the disadvantage of independent shipowners and traders, and in the end to obtain a practical monopoly of the traffic. In fact, private companies and traders in many cases would be compelled to withdraw from a business in which they would be subjected to hopeless competition.

Much was made of the argument that if powers were granted to the railway company the interests of the working classes would be prejudiced. The shipowners' petition continued :

There is a keen open competition . . . on the Clyde, and the result is beneficial to the public in keeping the fares at the very lowest figure, and providing cheap and convenient transit for the working classes and the general population of Glasgow, Govan, Partick, Renfrew, and Dumbarton to the various places on the coast. This the private owners could not continue to maintain unless they received a fair proportion of the traffic to and from the coast by train.

The point was elaborated by James Caldwell MP, who had 'made a careful study of the bill from the point of view of the working classes. . . . The mass of the people of Glasgow were much indebted to the river steamers for the health and enjoyment they had obtained through their means. He believed . . . that the river steamers would not pay if they were deprived of the railway traffic by the railway company putting on steamers of their own. The working classes would thus lose the benefit of the cheap fares . . . and consequently would not be able to enjoy the sea-breezes. . . . He looked upon the Clyde with its cheap steamers as the working man's highway to the sea air. The train and boat combined was the method of travelling of the middle and upper classes, who wanted to travel to and from the coast on business as expeditiously as possible. The question was whether Parliament should promote the interest of that class at the cost of the other. . . .' Mr Caldwell was supported in his views by John Wilson MP, who said that 'anything which was done to diminish the number of steamers running upon the upper reaches of the Clyde would be a great injury to the artisan population of Govan'. Councillor Crawford, of Glasgow, chairman of the Health Committee of the Town Council, thought it 'a very important matter that the artisan and poor classes of Glasgow, who lived in a very bad atmosphere, should have the maximum of possible opportunity of access to the river and estuary of the Clyde. . . . What he was anxious about was that if a monopoly service was developed by the railway companies it would necessarily stop the inducement of the private steamboat proprietors to sail from Glasgow, and that as a consequence of that evil results would follow to a population already in circumstances sufficiently bad as to health and fresh air. . . . The river at Glasgow was not at present in a tempting condition, but they were pushing forward the purification of the river. The view was that the Clyde was the highway of the people of Glasgow to fresh air, and they thought it should be in a condition fit to travel upon consistent with enjoyment.'

That the attitude of these witnesses might perhaps not be shared fully by the 'artisans and poorer classes of Glasgow' became plain a day or two later when a letter was published in *The Glasgow Herald* over the signature of 'A Working Man' who, in writing of the evidence given before Parliament, referred to 'the extreme prominence given by a number of these witnesses to what they are pleased to call the interests of the working classes. . . . As a working man I object to those gentlemen posing as the only guardians of working class interests. I am of opinion that even if Parliament pass this measure the working man's interest will be no more injured than they are. . . . It seems, so far as I am able to judge from the evidence, that their objections are solely in their own interests, and their attempt to enlist the sympathy of the committee by a show of regard for the interests of the working man is manifestly a blind. . . . I am surprised to

Page 71: Craigendoran boats: *(above)* the *Lucy Ashton* in the mid-nineties; *(below)* the *Lady Rowena* on the Arrochar service.

Page 72: Summer days: *(above)* the *Marchioness of Lorne*, with extended fore-saloon, approaching Dunoon; *(below)* the *Viceroy*, in final South Western condition, arriving at Cove.

find a temperance reformer like Mr Wilson, who is so earnest in his desire to remove all temptations to drinking out of the way, opposing what is undeniably a step in that direction. Surely Mr Wilson knows that when a man embarks on board a steamer at the Broomielaw, he embarks on board a floating public house, and if he is going, say, the length of Rothesay, that temptation is in his way for $3\frac{1}{2}$ hours; whereas, if he goes by rail and boat, that temptation is reduced to a minimum.'

The Glasgow & South Western Bill was opposed strongly by David MacBrayne. He was represented before the Parliamentary Select Committee by Alexander Paterson who said that the cream of the traffic on the Ardrishaig route was found before the Kyles of Bute were reached and that the *Columba* would therefore be vulnerable to railway competition. It was feared that Mr Mac-Brayne might lose the mail contract, which subsidised the winter service to a certain extent, but even allowing for the mail money it could not be carried on at a profit. The owners of the *Ivanhoe* were also represented by their purser, Mr Comrie, who drew the Committee's attention to the fact that the South Western would be in a position to compete with the *Ivanhoe* at every pier on her route and that it was feared that this would lead to her withdrawal from the Arran station. He said that his owners were willing to build a new steamer equal to the *Ivanhoe* to maintain Glasgow & South Western connections in winter as well as in summer between Ardrossan and Arran, and said that he saw nothing to complain of in the natural corollary that competing steamers from Ardrossan would then be under the management of Captain James Williamson—the one company would be perfectly distinct from the other !

Despite intense opposition to the Bill, two factors eventually swung opinion in its favour. The first was the obvious disadvantage under which the Glasgow & South Western Railway operated following the Caledonian onslaught and there was a feeling that the formation of the Caledonian Steam Packet Company had taken advantage of a loophole in the law to avoid Parliament's prohibition. Simple justice demanded that the South Western be granted equivalent facilities. Secondly, there was strong support for the company from the Greenock Harbour Trust and the Burgh of Greenock, amongst a number of other interests, including the Duke of Hamilton, the proprietor of the Island of Arran.

The evidence of these supporters won the day. On 3 July 1891 the Select Committee unanimously found the preamble of the Bill proved subject to the exclusion of Lochranza and the west of Arran from the area of the company's activities, and to the limitation of its service to the area west of Greenock, thus preventing its vessels sailing to Glasgow. There was also a stipulation requiring the company to charge similar fares to the Caledonian in respect of through bookings from Glasgow to Arran. The latter company's Ardrossan trains used the Barrhead and Kilmarnock Joint Railway for several miles, thereby incurring a toll from which the South Western, with an independent line of its own, was exempt, and it was thought that the potential advantage was unfair to the Caledonian. The exclusion of the west of Arran in addition to the ports on Loch Fyne and Kintyre was regarded as a sufficiently serious matter for the Glasgow & South Western promoters to recall Mr Murray, factor to the Duke of Hamilton, who said that Captain Buchanan had been in the habit of running the Ardrossan steamer to Lochranza and the west of Arran ferries at Pirnmill and Blackwaterfoot for several years, and that the service was so useful to the public that no pier dues had been charged. He feared that the exclusion of the South Western company would be an inconvenience to the inhabitants of Arran. Despite this additional evidence, however, the committee refused to change its mind.

So the Glasgow & South Western Railway obtained its powers to own and run steamboats. It had been a costly victory; witnesses' fees alone amounted to nearly £2,200 and in the end the limitations on operation were severe. Nevertheless, the task of recovering traffic lost to the Caledonian began at once. Arrangements had been made with Captain Buchanan's trustees[1] for the purchase of the *Scotia* and with the Lochgoil & Lochlong Company for

---

[1] Captain Buchanan had died in October 1890.

the *Chancellor* conditionally upon the passing of the Bill, and on 21 July 1891 payments of £7,500 and £6,500 respectively were authorised by the board in settlement of the purchase prices. Further arrangements were also made for the assignation of the Arran mail contract to the railway company by Buchanan's trustees. In the meantime Williamson's 'Turkish Fleet'—the *Viceroy*, *Sultan* and *Sultana*, and the *Marquis of Bute*—was purchased to form the nucleus of a fleet at Greenock. These purchases were agreed upon while the Bill was still before the Parliamentary Select Committee, giving rise to the following exchange :

> The Chairman—Captain Williamson is not a petitioner against the bill.
> Mr Balfour Browne (Counsel for the company)—Oh dear, no.
> The Chairman—In fact, he has been squared.
> Mr Balfour Browne—I would not use such a word; but he is not a petitioner against the bill.

Following the purchase of these steamers the directors proceeded to form a Steam Vessels Committee and advertised for a marine superintendent. On 25 August 1891 the following minute was recorded :

> The question of the appointment of a Marine Superintendent was considered, and, after interviewing three applicants, the Directors appointed Mr Alexander Williamson, who is to devote his whole time and attention to the Company's work, and receive a Salary at the rate of Four Hundred Pounds per annum—with promise of an advance soon, if he gives full satisfaction. The Engagement to be terminable at any time on three months' notice by either party.

Alexander Williamson was a son of the owner of the 'Turkish Fleet', who now retired from business. He had latterly been assistant to his father and responsible for radical improvements to the *Viceroy* early in 1891, so that he was evidently a man of practical ability as well as of administrative capacity. Not the least interesting feature of his appointment was that he was thereby thrown into direct opposition to his brother James, manager of the Caledonian Steam Packet Company.

STEAMERS OLD AND NEW

Alexander Williamson's responsibility of restoring the Glasgow & South Western position on the Clyde was demanding, for there could be no failure; instant success was essential. He set about the task with vigour, preparing a report for the directors on the condition of the stop-gap fleet which had been acquired and setting out the requirements for new steamers, and placed his findings before the board on 15 September.

Of the vessels purchased, the *Viceroy* was the most modern in her fittings following her extensive reconstruction in the spring of 1891. As built, she had been of the conventional design of the mid-seventies, with raised quarter deck, haystack boiler, and single diagonal machinery, a pattern that was dated by the late eighties. The alterations involved lengthening the steamer by fourteen feet and fitting her with proper deck saloons forward and aft of the engine room, while internally she was equipped with a dining saloon and smoking room. Having been reboilered as recently as 1888 she was therefore in thoroughly up to date order, although by no means as fast as the best Caledonian steamers.

The *Sultana* and *Marquis of Bute*, both built in the late sixties, were generally similar to the *Viceroy* as originally built, but neither had been substantially altered. They were distinctly behind the times, but both had a good turn of speed which compensated in part for their shortcomings. *Sultana* in particular had been a famous racer in the seventies and was always a favourite with the travelling public. The *Sultan* was the veteran of the fleet, dating back to the early sixties and in every respect was obsolete; she was therefore the first vessel to be disposed of on the arrival of new ships.

The other two steamers bought by the company were built in 1880, and were of quite different design. *Scotia*, as we have seen, was primarily intended for work in open waters all the year round, whereas the *Chancellor*, a vessel which closely resembled the more modern Loch Lomond steamers of her time, was built for cruising in the comparatively sheltered Loch Long. Like the

Viceroy, she was potentially a more useful vessel than most of the others, and underwent considerable alteration at the hands of her new owners.

The choice of colour scheme for the new fleet was inspired. After a brief experiment with black hull and red funnel with black top, tried on the *Chancellor*, a completely new livery, based on the Union-Castle liner colours, was adopted. Hulls were painted French grey, with white saloons, while the funnels were scarlet, with black tops. Paddleboxes were white, with elaborate painting and gilding. It was one of the most attractive colour schemes ever used on the Clyde, and added immeasurably to the appearance of even the plainest of steamers. The first vessel to appear in her new finery appears to have been the *Viceroy*, early in 1892.

The work of preparing specifications and drawings for new steamers was entrusted to Robert Morton, of Morton & William-son, who submitted proposals for two vessels at the end of September 1891. Tenders were invited on the basis of the specifications and on 13 October the Steam Vessels Committee agreed that contracts should be placed with J. & G. Thomson, Clyde-bank, for a steamer for the Arran service, and David Rowan & Son for a smaller vessel for the Rothesay station, Mr Morton being employed to superintend the building of both ships on the company's behalf. For the time being the question of ordering a second smaller steamer was deferred, but at the end of October the directors agreed to order a duplicate of the Rothesay steamer from David Rowan & Son, who sub-contracted the construction of the hulls to Napier, Shanks & Bell, of Yoker. The same builders and engineers had launched and completed the paddle steamer *Lorna Doone* to Morton's designs for Bristol Channel owners in May 1891 and this vessel was chosen as the model for the South Western steamers. Of proved design, with a good turn of speed, having achieved $17\frac{1}{2}$ knots, the *Lorna Doone* was a safe prototype. The Glasgow & South Western could risk no experiments and in consequence their new steamers differed in no major feature from the earlier one. The engines were of two cylinder, compound diagonal type with twin cranks, but the high pressure cylinder was made one inch greater in diameter than in the case of the *Lorna Doone*. There were two Navy boilers, with uptakes into one funnel, in a closed stokehold under forced draught. The prototype steamer was built with alleyways round her fore saloon, and her navigating bridge forward of the funnel but the South Western steamers' fore saloons were extended to the full width of the hulls, and the bridges were placed abaft the funnels to facilitate manœuvring at piers.

The new steamers received the names *Neptune* and *Mercury*. The former was launched on 10 March 1892 and ran her trials eighteen days later, achieving a speed of 18 knots. This was in excess of the guaranteed speed of 17.45 knots and the *Neptune* was immediately hailed as the fastest paddle steamer of her length afloat. On 31 March she went on an inaugural trip from Prince's Pier to the Kyles of Bute with a party of guests of the directors and the builders, dinner being served on the return run accompanied by the inevitable toasts—'Success to the *Neptune*', 'The Builders', 'The Superintendents', &c. *The Glasgow Herald's* account of the trip included a description of the steamer herself, from which it was evident that the standards of comfort set by the Caledonian steamers were to be followed in the Prince's Pier fleet:

> The decoration of the saloon is in the best possible taste. The seats, which are on the railway carriage principle, have spring cushions covered with the finest moquette; on the floor are pretty Brussels carpet runners; and dainty damask curtains adorn the windows. The first class dining saloon, which is seated for seventy and is situated below the general saloon, is furnished pretty much in the style of the drawing room. The ceiling and sides are beautifully decorated, and the floors laid with tastefully designed carpets and runners. On each side of the central passage are handsome mahogany tables, while right round the saloon run velvet-cushioned seats.

The builders' account for construction of the *Neptune*, amounting to £19,164, was authorised for payment on 12 April 1892 and on the following day the new steamer sailed down to Ardrossan to relieve the hard-pressed *Scotia* on the Arran service. That plucky vessel was brought in for heavy overhaul and reboilering at this time, the improvements including alterations to the ladies' cabin and saloon, and the provision of steam steering gear. She re-entered service during the second half of May as a relief steamer on the Arran station and inaugurated special excursions from Ayr. Meanwhile the *Neptune* and the *Duchess of Hamilton* were engaged in a hammer and tongs contest between Ardrossan and Brodick. The South Western service from Glasgow to Arran was accelerated on 2 May; an express train leaving St Enoch station at 9.5 am ran to Winton Pier, Ardrossan in 45 minutes and allowed passengers to reach Brodick at 10.40. It was stated that the *Neptune* did the crossing to Arran in an average time of 39 or 40 minutes, comparing favourably with the corresponding times by the *Duchess of Hamilton*, and representing a substantial improvement on the previous service. Indeed, the up afternoon service from Arran to Glasgow was reduced by more than half an hour after the accelerations.

The *Neptune* and *Duchess of Hamilton* were well matched. No longer could the Caledonian steamer leave Brodick five minutes after her rival and guarantee to be first into Ardrossan. Competition flared into open racing. On 14 May *The Glasgow Herald* published a letter over the pseudonym 'Prevention':

Now that the coasting season has fairly commenced, and with the great competition that exists this season among steamboat companies—greater than has ever been before—it is well, I think, that the travelling public should take a lesson in time, and be carefully on their guard as to safety. We all know the nasty mishaps that took place last year between Ardrossan and Brodick in one way and another, and I should not like to see a repetition of these before the season comes to a close. It may be gratifying to the respective companies to know that on a Monday morning lately the steamer belonging to A left Brodick Pier three minutes later than the steamer belonging to B, and, notwithstanding this, A steamer arrived at exactly the same time at Ardrossan as B steamer; but it is to be considered, on the other hand, that both steamers were running side by side with each other nearly all the way to Ardrossan at a high pressure of steam, having each a large complement of passengers, whose lives, I think, were somewhat endangered should anything have taken place. I do hope the proper authorities will see to this at once, and if need be alter the hour of sailings.

The writer's fears were not groundless. Seven days after the publication of his letter the *Neptune*'s reversing gear failed as she entered Ardrossan and she collided with the Caledonian pier, severely damaging her stem, but happily causing no serious injury to passengers. To add insult to injury, the South Western travellers on the outward journey had to complete their trip on board the *Duchess of Hamilton* while the *Neptune* was repaired.

Meanwhile, on 26 March 1892, a notable vessel had been launched at the yard of J. & G. Thomson, Clydebank. The new steamer, named the *Glen Sannox*, was intended as the regular vessel on the Arran route. She was substantially larger than the *Neptune* and cost no less than £30,000 to build. No expense was spared to produce a steamer which would recover the traffic lost to the Caledonian two seasons earlier. The *Glen Sannox* was indeed a masterpiece. She was a large, two-funnelled steamer, and was the first on the Clyde to have the plating carried up to promenade deck level at the bow, giving her a very modern appearance. Aesthetically she was superb, her two tall, elegant funnels, well raked, being in perfect proportion to the long grey hull. Her machinery was of the compound diagonal pattern, a very powerful set with impressively large moving parts. Steam was supplied by one double-ended and one Navy boiler, at a working pressure of 150 lb per sq in, an unusual arrangement, equivalent in capacity to three Navy boilers. She was as lavishly fitted out as her Caledonian rival but far outstripped her in the matter of speed. On trial on 1 June 1892 the *Glen Sannox* averaged $19\frac{1}{2}$ knots on two runs over the measured mile, achieving on one of these, doubtless with wind and tide in her favour, the amazing speed of $20\frac{1}{4}$ knots, while on four hours' continuous steaming round and about the firth she averaged 19 knots. These results confirmed the *Glen Sannox* as the swiftest steamer on the Clyde and the Glasgow & South Western Railway was at last in a position to reply effectively to the Caledonian challenge. Developments afloat were matched by improvements ashore for, simultaneously with the introduction of the new steamers, Winton Pier at Ardrossan was extended seawards by twenty feet and a new terminal station was constructed to cater for the Arran traffic.

The *Glen Sannox* took up her station on 6 June and on her arrival the Arran service was again accelerated. The 9.5 am express from St Enoch now gave a connection at Winton Pier allowing an arrival in Brodick in eighty-eight minutes from Glasgow as compared with the former ninety-five, but the fastest service was provided by the 1.55 pm down train on Saturdays and the 4.55 pm (except Saturdays), both of which allowed an eighty-five minute journey to Brodick, 105 minutes to Lamlash, 115 to King's Cross and 125 to Whiting Bay. But the fastest timing of all was the 8.5 am from Brodick every day, which landed passengers in St Enoch station, Glasgow, in precisely eighty minutes. Like her Caledonian counterpart, the *Glen Sannox* gave a programme of daily excursions between morning and afternoon railway connections. On Mondays and Wednesdays she cruised round Arran, on Tuesdays and Thursdays to Ailsa Craig, and on Fridays to Campbeltown Loch where, according to the strict interpretation of the company's powers, she was entitled to go so long as no call was made at Campbeltown itself. Under the command of Captain Colin McGregor, himself a native of Arran, the *Glen Sannox* became an immediate favourite.

The *Mercury*, the second of the two smaller steamers being built for the company, was launched on 18 April 1892. The vessel ran her trials on 18 May in stormy conditions but maintained 18.45 knots as the mean of two runs over the Skelmorlie mile, an even better performance than had been achieved by her sister ship. The company was now in the fortunate position of having three very fast, modern steamers available for the 1892 summer season, while of the second-hand purchases of 1891, four ships had been substantially modernised within the previous two years. During the winter of 1891-2 the *Marquis of Bute* had undergone reboilering at the hands of King & Co, Glasgow, at a cost of £1,000, while the *Chancellor* was even more thoroughly improved by the replacement of her boiler, the substitution of a surface condenser for her original jet condenser, and the conversion of her double diagonal machinery to compound expansion.

RAILWAY PROBLEMS

The tremendous improvements in the fleet were accompanied by the introduction of new passenger carriages on the coast trains to Prince's Pier at the beginning of May 1892. They were the first bogie vehicles to be used on these services, and were built by the Metropolitan Railway Carriage & Wagon Company, Birmingham, and the Birmingham Railway Carriage & Wagon Company. In due course their use was extended to the Ardrossan section. There were three types of carriage—first class, with six compartments; third class, with seven compartments; and composite, with three first and four third class compartments. 'The first class compartments are exquisitely panelled in polished sycamore, banded with bird's-eye maple, and trimmed in blue, with crimson blinds. In the third class compartments also the interiors are decorated in a very pleasing and effective style, and the cushioned seats are perhaps even more luxurious than those in the older first class carriages.' Unfortunately for the company the increase in weight of the new boat trains came as something of an embarrassment to the competent, but relatively small, Smellie 4-4-0s, the 'Wee Bogies', which then ran the expresses. The gradient profile of the Prince's Pier line between Paisley and Greenock can best be likened, in the words of a prominent authority, to the gable of a house. The line climbs sharply from Cart Junction, west of Elderslie, to its summit beyond Kilmacolm before descending steeply into Greenock. Scenically the route was superb but from the operating standpoint it was far from easy.

The fast South Western timings achieved in the years before the Great War were made by long, slow climbs followed by hair-raising descents to Prince's Pier. The Smellie engines could run all right, but they simply did not have the power to pull fast uphill. James Manson, who succeeded as locomotive superintendent of the company in 1890, turned his attention to the production of a larger design but this class, the famous 'Greenock Bogies', did not appear until 1895; thereafter they practically monopolised the route until just before the outbreak of war in 1914. The lack of power of the Smellie engines caused adverse comment. A newspaper correspondent wrote :

> If the company wish to win back the traffic which they lost three years ago . . . they won't do so by a laggard and sleepy hollow system. New boats and new carriages are all very well in their way, but if the journey from station to station is not accelerated the advantage over the old style is of very little account. . . . There is no blinking the fact that the public will travel by the route which will bring them fastest to their destination.

Although conceding the difficulties of operating the Prince's Pier line, another writer said that

> . . . the South West might do better and please those who favour it if, on the express runs at any rate, the company would afford an extra engine for the up-gradients that are the cause of considerable delay.

Nevertheless, these were only teething troubles. The public, despite them, was coming back to the Glasgow & South Western in no uncertain manner. The weather at the Glasgow Spring Holiday of 1892 was brilliantly fine and *The Glasgow Herald* reported that 'the demand from St Enoch Station to Greenock could only be supplied by seven specials'. The Glasgow Fair Holiday in July was also exceptionally fine and again traffic surged to exceed all previous records at St Enoch. 'Besides the ordinary trains to Greenock, it was found necessary to despatch four specials in order to overtake the heavy traffic.' The passengers by these were, of course, mostly bound for the coast towns connected by steamer with Prince's Pier. There was a marked increase in the Arran traffic, many excursionists being naturally desirous of testing the fine seagoing qualities of the *Glen Sannox*.

The success of the South Western service in attracting back traffic was undoubted, but the cost had been heavy, and at the company's half yearly meeting of shareholders in Glasgow on 13 September 1892 the chairman, now Sir William Renny Watson, was challenged on this matter. He said that the working expenses of the steamboats, depreciation and insurance amounted to nearly £23,700 but that there was ascribed to the steamers a large portion of the increase in the general passenger receipts. A shareholder said that the steamers were excellent in their way but the chairman had not told the meeting what the amount of the receipts from the steamers had been. The somewhat evasive reply to this questioner was that it was difficult to apportion truly and justly the receipts of the steamers and show them in a separate account. Mr Brown (the shareholder) must trust the directors to see that these receipts were as large as possible, and also that they were properly dealt with. If the shareholders were not pleased with the result, the remedy lay in their own hands. In brief, the long-suffering shareholders were being given the broadest of hints to drop a matter which the board almost certainly knew could cause much embarrassment if brought into the open. Surviving Caledonian records of that period indicate that most of the vessels in the fleet were running at a loss and there is little reason to think that the Glasgow & South Western Railway was in any better case. Writing to *The Glasgow Herald* a day or two after the meeting, a Mr Logan stated that a careful study of the coast traffic suggested to him that for every twenty shillings spent on maintenance of the steamers, the company received only seven in revenue. There is no way of checking his calculations, but Mr Logan cannot have been far out. The truth of the matter was that the Caledonian and South Western companies were being drawn into a ruinous competition for the coast traffic. Thus far, the North British had stood aloof but in due time this company too would be drawn into the battle, and only the passing of time and spiralling costs would bring the contestants to their senses. It is true that there were attempts to prevent the thing getting out of hand; the South Western minutes of 13 Sepember 1892 record an approach by the Caledonian Company with a proposal that the Ardrossan—Arran winter service should be worked for alternate periods by the two companies to keep down costs, but the South Western declined to agree, offering instead to recognise Caledonian tickets and carry their passengers on receipt of the steamboat portion of the through fares. This in turn was unacceptable to the Caledonian. Nevertheless, a tripartite series of meetings was held between representatives of the three railways early in 1893 with a view to arranging a scheme for lessening the costs of running the coast traffic. Eventually the North British withdrew, but the Caledonian and South Western companies agreed to take off one steamer each during the ensuing summer season and to honour each other's tickets between Glasgow, Paisley and the coast.

THE NEW PRINCE'S PIER

It was a small concession to sanity and ultimately had little effect, for competition flared up again in 1893 and once more the Glasgow & South Western Railway found itself spending heavily in the provision of improved facilities for its Clyde coast traffic. This time the station and ancillary buildings at Prince's Pier were rebuilt and extended. There was reported in May 1892 the acquisition of additional ground alongside the old station at a cost of 'fully £42,000' and a Traffic Committee minute of 6 June 1893 recorded that 'an amended plan of the proposed alterations at this Station estimated to cost £20,750, of which £5,400 is chargeable to Revenue was submitted, and it was agreed to recommend the Board to approve of the Plan'.

More money was spent on the steamers. The subject of new tonnage was under review as early as October 1892 and on 10 January 1893 the Steam Vessels Committee met and 'the question of providing new steamers for next season was again considered, and designs for a Boat of 200 feet in length prepared by Messrs J. & G. Thomson (Limited) were submitted. Resolved to recommend the Board to build two new steamers, in the event of the sale of the *Scotia* being satisfactorily carried through, and to build one steamer only, if the *Scotia* is not sold.'

Determined efforts were made to sell the *Scotia* at this time but due to a series of misfortunes she was not disposed of until late in 1893. It was minuted on 10 January that the general manager had arranged to sell her to Messrs Thomas McLaren, of Glasgow, for £10,500, but at the end of the month it was reported that the prospective purchaser objected to the vessel's coal consumption and the sale fell through. She was then offered for sale again and on 9 May it was stated that an agreement had been entered into with a M. Jean Thevenet. On 6 June the purchaser was reported to have asked for a modification of the conditions of sale allowing him to settle the purchase price over a period of three months, the steamer to be chartered in the meantime, but this arrangement also fell through despite the railway company's agreement. The general manager was thereupon instructed to advertise the *Scotia* for sale again, which he did without success, and on 18 July authority was given for the upset price to be reduced to £6,000. Still no offers were received and the vessel was again advertised at the further reduced price of £5,000 on 23 August. There was still no luck and the upset price was reduced for a third time, to £4,500 at which she was exposed for sale by public roup on 13 September, when she was at last sold, for £5,300, to Mr Frederick Edwards, of Cardiff.

In the meantime the board had authorised the ordering of the second new steamer referred to in the minute of 10 January 1893, provided that J. & G. Thomson, Limited, could guarantee delivery not later than 25 June. This was agreed and Robert Morton was again retained to supervise the construction of the new steamers. On 11 April the minutes recorded the signing of a formal contract and 'it was resolved that the New Steamers now being built, be named "Minerva" and "Glen Rosa".' The pattern of nomenclature for the new South Western ships continued to favour Roman deities, with deviations only in the case of steamers intended to have a close connection with the Arran services: the *Glen Rosa* was designed to undertake the winter sailings from Ardrossan.

The new vessels were sister-ships, rather smaller than the *Neptune* and *Mercury*, but sturdily constructed for all the year round service. Alongside them at Clydebank a third vessel, of identical design, was also built. She was the *Slieve Donard*, for the Belfast & County Down Railway, better known in later years as P. & A. Campbell's *Albion* on the south coast of England. All three ships shared an unusual feature, unique as far as Clyde steamers were concerned. The deck at the bow was raised to the level of the mainrail and an open rail fitted above it. The intention was doubtless to avoid the risk of undue flooding in heavy seas.

The *Minerva* was launched on 6 May 1893 and Robert Morton stated that the builders hoped to deliver the steamer by 1 June, while in the case of the *Glen Rosa* work was well advanced. The latter vessel was launched on 31 May by Miss Guthrie, daughter of the deputy chairman of the company. By 20 June Morton was able to report that rapid progress was being made with the *Glen Rosa* and that it was expected that she would be delivered by 1 July. In the event her trials took place on 27 June when a mean speed of $17\frac{3}{4}$ knots was achieved over the measured mile, and the contract price of £17,500 was authorised to be paid by minute of 4 July.

The Glasgow Herald reported that the *Glen Rosa* and *Minerva* had been specially designed not only for service in winter but also for summer excursions in the outer waters of the firth, the Board of Trade certificate enabling the *Glen Rosa* to ply as far as Campbeltown and Stranraer. The engines were of conventional compound diagonal pattern, supplied with steam from a double-ended boiler, with uptakes into a single, smartly-raked funnel. The navigating bridge, as in earlier steamers, was situated on the combined ticket office and captain's cabin on the promenade deck abaft the funnel. Like the company's other new ships, they were well finished :

Two broad stairways lead to the main deck, at the after end of which is a spacious saloon for first-class passengers. This department is panelled in polished oak, and the upholstery and decorations being in harmony, a rich and striking effect is produced. There is a comfortable dining saloon below the main saloon, with large pantry and bar, and at the forward end of the ship the accommodation for second-class passengers as well as for the officers and crew is exceptionally comfortable.

The first recorded service of the *Glen Rosa* was on Saturday, 1 July 1893, when she was advertised to take the additional run 'for the convenience of FAMILIES REMOVING to ARRAN', sailing from Ardrossan in connection with the 12.35 train from Glasgow and calling at Corrie, Brodick, Lamlash, King's Cross and Whiting Bay. This being essentially a slow run intended for the carriage of large quantities of luggage, the *Glen Sannox* was released for the normal fast service—'*No luggage will be conveyed*'.

In spite of the addition of two new ships, numerically the fleet remained unaltered at the end of the 1893 summer, for not only had the *Scotia* been sold to the Bristol Channel but the old *Sultan* had also been disposed of, in this instance to Captain John Williamson, who purchased her for £500 in April, he having set up in the steamboat business on his own account in succession to his father, Alexander Williamson, Sr. It was a modest start, destined to lead to much greater things.

Expenditure on the older ships was to be expected, but it must have come as a surprise to the board to be informed of defects in the machinery of the *Neptune* and *Mercury* as early as August 1892. Robert Morton was consulted and recommended strengthening of the engine framing, which was carried out and no further trouble seems to have been experienced. The precise nature of the defects cannot be ascertained, but it is possible that the engine entablatures were lightly made for powerful vessels regularly run at full speed. They were of the older pattern, four in all, identical in size, rather than the one large and two smaller entablatures of more modern compound engines. The repairs

Page 89 : South Western beauty : the *Jupiter* in Rothesay Bay.

Page 90: Turbine power: the *King Edward* on trials on the Skelmorlie mile in 1901.

were carried out by the original engineers, Messrs Rowan, at a cost of around £1,000.

The season of 1893 was a mixed one for the company in many respects. The reconstruction of Prince's Pier was in full swing, inhibiting to a degree the smooth flow of traffic at this important railhead, and there were still faults in the train service causing 'Brandane' to refer to the subject in the press early in July. 'I have waited,' he wrote, 'in the hope that some abler pen than mine would take up the matter of the G. and S.W. Railway service of trains to the coast . . . and although several letters have appeared complaining of the trains on other sections of the line, never a word has been written regarding the faults of the present service via Greenock. A month has elapsed since the new express trains were inaugurated from St Enoch to run in connection with the new swift steamers. It was reasonable to suppose that the officials would have remedied any defects which might have come to their knowledge, but, whether in defiance of public convenience or indifferent to their company's reputation, no improvement has taken place. Since the beginning of June I have travelled almost daily by their 4.20 pm to Rothesay during the week, and on Saturday either by the 2.3 or 4.20. I do not think that on any one occasion did these trains start to advertised time—the late time ranging from five to 15 minutes. Sometimes the carriages, particularly on the Saturdays, did not arrive at the platform till much after the time of starting, and then it was a rush and scramble to obtain seats. The evil did not end there. Instead of sending these expresses via the comparatively quiet Canal line to Paisley, the old line (Glasgow & Paisley Joint) was requisitioned as of old, and it was invariably the case that the signals remained against us from Shields Road to Paisley. . . . In consequence it was something to be proud of if we reached Prince's Pier in 50 minutes, instead of the advertised 38. The boats, of course, were thrown correspondingly late, and the Caledonian steamers got the piers first. If the Glasgow and South Western wish to win back a fair portion of their lost coast traffic they must show some vim in the despatch and time-keeping of their trains. What is the good of possessing new engines, new carriages, and a magnificent fleet of new steamers if they cannot keep faith with the public?'

Brandane's rebuke seemed to have been taken to heart and gradually the troubles of which he wrote were eliminated. There was nothing fundamentally wrong with the coast service and in time the traffic built up satisfactorily again after the appalling slump of the late eighties. In May 1894 the new Prince's Pier railhead was formally opened, completing the last stage of the modernisation of the coast service. A special train carrying officers of the company and guests, together with Press representatives, left St Enoch on the morning of Friday 25th and ran to Greenock, where an inspection was made of the new station and pier buildings, and thereafter the party cruised to Loch Striven in the *Neptune*. The new buildings were the most imposing of their type on the Clyde. Although substantially on the site of the old station, the new one was larger than its predecessor. Four Italianate towers dominated the wide façade, constructed of red Ruabon brick. The platforms, protected by verandah roofs, were above pier level and at right angles to it, and two carriage ways, two flights of stairs and two luggage lifts connected upper and lower levels. The upper part housed Captain Williamson and his staff as well as railway officials. The plans and specifications of the new station were prepared by Mr Melville, engineer to the Glasgow & South Western Railway.

RACING DAYS

Competition between the South Western and Caledonian companies reached a new intensity in the summer of 1894. There was no doubt that risks were taken which might well have led to serious accidents both ashore and afloat. A letter published in *The Glasgow Herald* on 9 June sought to draw official attention to the risks inherent in the thunderbolt descents to Prince's Pier by the fastest expresses. Hugh Smellie's small-wheeled 4-4-0s still ran most of the trains but the large-wheeled Manson 4-4-0s of the

No 8 Class, introduced in 1892, now supplemented them, with electrifying results:

> I would direct the attention of the G. and S.-W. officials to the dreadful speed adopted by their express trains down the hill to Greenock,

wrote the alarmed correspondent.

> On Friday evening the 5.5 express had a narrow escape of leaving the metals just beyond Lynedoch Station, where there is a sharp curve. As the tickets are lifted at Paisley, the trains do not require to stop at Lynedoch Station, which has hitherto been the safe point of the hill; consequently the curve in question is negotiated at full speed, and a few repetitions of Friday evening's performance will only have one end.

Once on board the steamer the passenger was still exposed to the dangers of racing. On Tuesday, 17 July, the *Minerva* had just left Dunoon pier, then in its earlier, smaller condition prior to reconstruction, and the old Buchanan steamer *Vivid* was signalled to follow her into the empty berth. The Caledonian vessel *Galatea*, however, in hot pursuit of the *Minerva*, cut in ahead of the *Vivid* and the resulting row ended in the Marine Police Court, Glasgow, where Captain Duncan Bell of the *Galatea* was fined three guineas for his escapade. Undaunted, he returned to the fray and on 7 August the same rivals met off Kirn, both in connection with fast expresses from Glasgow. The Kirn piermaster decided that the South Western steamer was in a more favourable position and signalled her in first, but the indefatigable Captain Bell refused to concede defeat and came in ahead of the *Minerva*, which had to reverse but could not avoid a collision. Again there was an appearance at the Marine Police Court, followed by Captain Bell's conviction. Witnesses giving evidence at the trial spoke of the partisanship engendered by the railway rivalry and there were dark accusations of discrimination in favour of one or other company by piermasters and their assistants.

It was, and remained, the custom for Clyde steamers to be chartered to private parties for excursions in this period and later years, and on 22 June 1894 the *Glen Sannox* was so employed to convey the Trades' House of Glasgow on their annual trip. A special train with a party of three hundred people left St Enoch for Prince's Pier at 10.10 am, and the steamer sailed thence to Dunoon and Innellan, Largs, Millport and Lamlash, and continued to make a rare call at Skipness, a remote pier on the Kintyre peninsula. Here the party went ashore for an official photograph to be taken and then, as the *Glen Sannox* sailed back through the Kyles of Bute, dinner was served. The steamer reached Toward Point in time to meet several warships of the Channel Fleet steaming up the firth on their way to anchor off the Tail of the Bank and the passengers, evidently in mellow, post-prandial mood, 'were at once on the *qui vive*, and as the *Glen Sannox* passed them her flags were dipped, handkerchiefs were waved vigorously, and "Rule Britannia" was sung by one and all several times, accompanied by the band which had given selections during the sail. These acts of patriotism and respect were generously recognised by the commanders of the different vessels, which in turn dipped their flags, and then the *Glen Sannox* went direct to Greenock. . . .'

With this period piece we may conveniently leave the Glasgow & South Western. The improvement wrought in a few years was immense and owed much to the ability and initiative of Captain Alexander Williamson. With rather less publicity, he was doing as good a job for his company as James Williamson was at Gourock. His directors thought well of him, and in August 1894 his salary was raised from £400 to £450 per annum.

In five short seasons the Clyde coast scene had been utterly transformed by the competing railway rivals on the Renfrewshire and Ayrshire coasts, and it is now necessary to retrace our steps to 1889 to follow the progress of the North British Steam Packet Company and the private owners in this period of rapid change.

CHAPTER FOUR

UP RIVER AND DOWN FIRTH
(1889-1894)

WHILE the Caledonian and Glasgow & South Western companies fought for the steamer traffic from the southern shores of the firth during the early nineties the North British Steam Packet Company quietly modernised its fleet sailing out of Craigendoran, and the Inveraray Company replaced the *Lord of the Isles* with a larger and faster ship of the same name. The Buchanan fleet, on the other hand, suffered retrenchment as older steamers were sold or broken up, and only one new vessel was added. The old Wemyss Bay and 'Turkish' fleets disappeared from the Clyde, although in some cases the individual members continued to sail under other flags. A notable example was the *Victoria* which returned to the firth after a short absence and took up service to Campbeltown during this period.

THE CRAIGENDORAN ROUTE

The *Jeanie Deans* was a notable racer, and when it was announced at the start of the 1889 summer season that the North British flagship had been fitted with a surface condenser and improved so that she was sailing faster than ever, her admirers looked forward to some memorable jousts with her rivals. Even the *Columba* found the *Jeanie* an adversary to be reckoned with; in the previous summer Robert Darling wrote to David Mac-Brayne to say that his steamer was sailing to Ardrishaig with a special excursion and that she would probably encounter the *Columba* in the Kyles of Bute and, in plain language, what about a race? The two vessels met as they approached Ardlamont and the *Jeanie*, keeping to the south of the MacBrayne steamer, overhauled her halfway across Loch Fyne, crossed her bows, and continued her cruise. Another contest between the two was observed on 31 August 1889, on this occasion between Rothesay and Innellan. The *Jeanie Deans* was due to leave Rothesay at 3.50 pm and call at Craigmore before Innellan. The *Columba*, behind time due to holiday traffic, left Rothesay at 3.53 pm and by the time the North British steamer cast off from Craigmore the *Columba* was hard on her heels, about four lengths astern. The observer of this race reported that both steamers were coaling up vigorously and that on rounding Toward Point the *Jeanie* was keeping her distance. From these two incidents and from other reports that have survived it may be inferred that the North British Steam Packet Company was fortunate in possessing an unusually speedy vessel, possibly the fastest of her period. Poor *Jeanie*! A greyhound in her youth, she was destined to enjoy little over a decade of fame before changing patterns of comfort and public demand forced her rebuilding into a saloon steamer in the middle nineties. She was never the same again, her speed thereafter being very ordinary by the standards of later years. It was her fate to become one of the most notorious of Clyde steamers by virtue of her Sabbath-breaking activities over several seasons. Latterly she became a sober and respectable member of the Buchanan fleet, in which condition she was best known to a later generation, but in her heyday we do well to recall her as one of the crack vessels on the river.

Another North British steamer, the *Diana Vernon*, was equipped with a surface condenser in 1889. This small, neat vessel was normally seen on the Craigendoran–Dunoon and Craigendoran–Gareloch routes, and while on an early run down from Garelochhead on 6 August she had the bad fortune to collide with a floating log, smashing one of her floats. She was unable to continue her sailing and had to lie at Row pier until the *Gareloch* took over, her passengers in the meantime being taken to Craigendoran by Buchanan's veteran *Balmoral*.

Caledonian developments in 1889 were countered in a small way by the introduction of a brand new express train for the

North British coast service between Craigendoran and Glasgow. Introduced in June, this was made up of seven 50 ft bogie carriages, of which no fewer than five were first class. These had seven compartments each, while the two third class vehicles which made up the train had eight compartments. There was no provision for second class which, it will be remembered, had been abolished earlier in the same year. The new North British train was lit by gas, but it was supplemented by an ingenious electric lighting system which drew current from conductor rails within the tunnels on the Glasgow underground section, thus avoiding the necessity of fitting the coaches with batteries or dynamos.

WEMYSS BAY FINALE

Captain Alexander Campbell's Wemyss Bay fleet of steamers was much in the news during 1889. The flagship, the *Victoria*, was a modern, powerful ship and in her day was a force to be reckoned with; at this period she was timed to connect with the 4.35 pm express train from Glasgow (Bridge Street), the service on which her celebrated predecessor the *Sheila* (later *Guy Mannering*) had made a great reputation. Some fast runs were made early in June, culminating in a record on the afternoon of the 10th, when the *Victoria* reached Rothesay at 5.54 pm. The journey from Glasgow, much of it over single line (the Wemyss Bay Railway had not then been doubled) and including the change from train to steamer, had occupied only seventy-nine minutes. Partly because of exploits such as these, and also on account of her superior passenger accommodation, the *Victoria* was much in demand for excursions, a traffic which Captain Campbell not unnaturally tended to favour at the expense of railway connections, since financially the latter were less lucrative. Thus we read of the *Victoria* as being much sought after as club steamer during the Clyde Fortnight of 1889. It was Captain Campbell's misfortune that special charters of this kind, although profitable, nevertheless built up trouble for the future. The remainder of his fleet were individually much inferior to the *Victoria* and caused much adverse criticism when they deputised for her on the regular runs. Complaints from the public inevitably recoiled on the Wemyss Bay route as a whole so that recriminations were not infrequent between the Caledonian Railway directors and Captain Campbell. There had, indeed, been an occasion several years earlier when the railway company's representations had led to the placing of the *Victoria* on the through services with a view to effecting an improvement in standards, but there had been earlier disagreements, including a withdrawal of the steamers for a short time and between one thing and another matters were rapidly approaching a final confrontation. Captain Campbell's basic complaint was that he received too small a fraction of the through fares from Glasgow and in the circumstances it was natural that he should wish to take advantage as much as possible of charters for his largest and most costly vessel.

We have already noted the appearance of another Wemyss Bay steamer, the *Adela*, in the newspapers of 1889, under less happy circumstances, when she ran down a small sailing boat between Toward Point and Craigmore, causing the drowning of a young doctor. The tragedy occurred on the evening of 13 August when the *Adela* was on her last down run to Rothesay. She was steered from the navigating bridge abaft the funnel, in common with the vast majority of Clyde steamers of her time and John Glen, the pilot, failed to see the boat as it lay in the steamer's path, slightly on her starboard bow. Glen was steering from a position on the port side of the wheel and his view was obstructed by the funnel. By an unhappy chance, Captain Archibald MacLachlan, her master, had just gone below, otherwise an accident might well have been avoided. As it happened, Glen's first hint of trouble was a warning shout from a passenger. The vessel was hove to instantly, but too late to avoid the sailing boat with its two occupants from being smashed by the starboard paddle wheel. A boat was lowered at once and one of the men was rescued but the other disappeared and his body was never found. The disaster led to much criticism of the practice of steering from abaft the funnel but much of it was ill-informed.

Properly manned, the steamers of those years could be quite safely navigated in this way; furthermore a bridge between the paddleboxes rather than forward of the funnel had many advantages in the important matter of taking piers. Both captain and pilot of the *Adela* were arrested and charged as a result of the accident but were acquitted, the court holding that they had not been guilty of negligence.

Bad feeling between Captain Campbell and the Caledonian Railway came to a climax in the spring of 1890. Despite his obligation to hold to the terms of his contract until the autumn, Captain Campbell abruptly gave notice in April that he proposed to withdraw the steamers at the end of the month, possibly hoping that the railway company could be forced into granting better terms, at least on a short-term basis, for it was obvious that it was only a matter of time before the Caledonian Steam Packet Company would have enough tonnage to undertake the Wemyss Bay service on its own account. The railway board refused to be coerced and made arrangements to take over as soon as the Campbell steamers were taken out of service. There was no serious trouble; the Caledonian ships took up the runs at the beginning of May and the days of the white funnel fleet at Wemyss Bay ended overnight, after a quarter of a century. Campbell's vessels were sold off the river almost immediately, the *Victoria* being sent to Belfast Lough. The little *Lancelot* lingered on the Clyde long enough to be noted on a very unusual sailing from Paisley to Dunoon, Largs, Millport and Rothesay on 14 August. The River Cart was then being dredged and widened for passenger and other navigation and by the summer of 1890 a special trip of this kind by a small steamer was possible, although the scheme as a whole was not finished until the following year.

THE LORD OF THE ISLES

More important changes were imminent in 1890 but most owners waited to see the effect of the Caledonian expansion. It came as an unexpected surprise when, in mid-September, it was announced that one of the most famous pleasure steamers on the firth, the *Lord of the Isles*, had been sold for service on the Thames, although the shock was tempered by a simultaneous announcement that a successor was to be built to carry on the name and tradition. Many people must have queried the wisdom of the move, for the *Lord of the Isles* seemed to be as good as ever and she was only a year older than her rival, MacBrayne's *Columba*. Together these large steamers had become household words on the Loch Fyne service and their reputation extended far beyond Scotland. In normal circumstances there could have been little incentive to dispose of an important steamer of the standard of the *Lord of the Isles*, but the Caledonian company was setting entirely new standards by 1890 and the Glasgow & Inveraray Steamboat Company's directors may well have thought that the cost of an extensive modernisation and refitting of their vessel would be better employed in the construction of an entirely new steamer.

It was a common practice in those days for tourist steamers to start the season with an opening cruise on which the owners entertained their friends and guests before starting the regular summer service. The last occasion of this kind involving the *Lord of the Isles* took place on a day of mist and rain in May 1890. Sailing from the Broomielaw the steamer called at Prince's Pier and Gourock to pick up passengers before cruising past Largs, between the Cumbraes and via Rothesay Bay to Loch Long where, it was recorded, 'the dinner gong was sounded and the guests tripped down to the saloon where . . . a sumptuous dinner was served and thoroughly enjoyed'. It was observed that the saloon had been completely changed and was now handsomely fitted up as a drawing room, complete with piano and an American organ, tuned to the same pitch so that both instruments could be played together. Special attention was drawn to the arrangements for a weekend trip to Inveraray. For the sum of thirty shillings the passenger was conveyed from Glasgow to Inveraray and back, enjoying two nights and a day in a local hotel, including all meals ashore and afloat, a bargain which the company continued to advertise for many years.

The Lochgoil & Lochlong Steamboat Company, Ltd was closely associated with the Inveraray company under the managership of M. T. Clark, and the same season of 1890 which saw the departure of the *Lord of the Isles* was the last full year in which the *Chancellor* sailed as a member of the former company's fleet. She was sold to the Glasgow & South Western Railway early in 1891 together with the Arrochar connection which she had maintained for eleven years, originally as a member of the Lochlong & Lochlomond fleet. The two remaining steamers of the Lochgoil & Lochlong Company were thereafter retained mainly on the Lochgoilhead route and the regular Arrochar sailings in connection with the Loch Lomond tour abandoned to other operators. Foremost among these was the North British Steam Packet Company which of necessity had to assume responsibility for the service which had been run from Craigendoran pier so long by the *Chancellor*. A new steamer was ordered from Hutson & Corbett, Glasgow, to be ready for the summer of 1891. Boiler and engine were provided by Hutson & Corbett but the building of the hull was sub-contracted to S. McKnight & Co, of Ayr.

NEW STEAMERS AT CRAIGENDORAN

There was something of a revival in North British fortunes in 1891 for in addition to the new Arrochar venture, a modern replacement for the steamer *Gareloch* was ordered, also from Hutson & Corbett, although in this instance the construction of the hull was undertaken by J. McArthur & Co, of Paisley. Although the *Gareloch* was the oldest member of the Craigendoran fleet she was still only eighteen years old and there was useful work for her elsewhere. Accordingly she was sent to the Firth of Forth to join the fleet of the Galloway Saloon Steam Packet Company Ltd, a subsidiary of the North British Railway. Arriving at Leith on 25 June 1891 she was renamed *Wemyss Castle* and took up sailings between Leith and Fife ports, a service on which she remained for the rest of her career, being ultimately withdrawn and broken up in 1906.

The new steamer for the Arrochar service was launched on 31 May 1891 and received the name of *Lady Rowena*. She was a saloon steamer, ten feet shorter than the *Jeanie Deans*, but no attempt was made to produce an outstandingly fast vessel. The emphasis was rather on comfort in view of the main service for which she was built. The fore-saloon was extended forward of the mast and fitted out as a spacious dining room. It would have been even roomier had not the old-fashioned arrangement of main-deck alleyways round the saloon been perpetuated. The main saloon aft, however, was made the full width of the hull. The standards of furnishing were high. A contemporary newspaper account described the saloon as being 'handsomely fitted up in polished hardwood, with gilt cornices, having large plate-glass windows all round; artistic Lincrusta ceiling, tinted and relieved with gold mouldings; sofas all round upholstered in Utrecht velvet, Brussels carpet and runners on floor, and windows neatly draped with damask curtains'. There were some unusual features in the new boat, amongst which may be instanced the lack of any passenger accommodation aft below main deck level, the quarter deck raised to main rail level abaft the main saloon, and the placing of the steering wheel forward of the funnel on a raised platform, although the navigating bridge remained between the paddleboxes. In this arrangement, of course, the *Lady Rowena* was not unique, for the Lochgoil & Lochlong Company's *Windsor Castle* was similar.

In the engine room the *Lady Rowena* conformed to well-established practice. She had a haystack boiler giving steam at 50 lb per sq in to a single diagonal engine which had the modern refinements of steam starting gear to move the main piston past 'dead centre' position, and a surface condenser instead of the older jet type. Steamers with the latter were being generally rebuilt at this time and of the North British steamers the *Guy Mannering* had been so altered early in the previous year following the successful conversion of the *Jeanie Deans* in 1889. The

Lady Rowena was ready by the middle of July 1891 and ran a preliminary speed trial over the Skelmorlie mile while on her way up-river from Ayr on the 14th; no record is available of her performance although it was stated that the result was 'very favourable'. The *Lady Rowena* was evidently regarded as the showpiece of the North British Steam Packet Company's fleet, for Captain McKinlay of the *Jeanie Deans* was transferred to her together with Mr Minto, the purser, and Mr Angus, the engineer. There was an innovation for the company in the employment of a stewardess and this, together with the importance attached to the accommodation generally, indicated the intention of the company to rely on comfort rather than speed on this predominantly tourist service.

The other North British steamer of 1891 was launched jointly by the Misses Darling, daughters of the company's general manager, on 24 June and named by them *Lady Clare*. This was a smaller vessel, no longer in fact than her predecessor, but she was slightly beamier and had full deck saloons fore and aft. Being intended for general service runs on the Gareloch route, she was not as lavishly furnished as the *Lady Rowena* but was a distinct advance on the *Gareloch*. In general, the *Lady Clare* conformed externally to North British standards, with bridge abaft the funnel and in common with the other steamers she was driven by a single diagonal engine and had a haystack boiler.

A NEW LORD OF THE ISLES

The most important steamer to enter service in 1891 was the new *Lord of the Isles*. Like her predecessor she was designed and built by D. & W. Henderson of Meadowside, Partick, and in all essentials the new ship could be described as an improved version of the old. She was nine feet longer and slightly broader than the first steamer, and resembled her in appearance, having two funnels fore and aft of the paddleboxes, but the deck saloons were made the full width of the hull, affording greater covered accommodation. The new vessel was fitted with the same pattern of diagonal oscillating machinery as in the first *Lord of the Isles*; this was a rare type on the Clyde and its application to the Inveraray steamers and the *Ivanhoe* suggests that it was a Henderson patent. There were two cylinders, fore and aft of the crankshaft, driving upwards on to a common crank. It may be questioned whether the owners were wise to perpetuate a form of engine which was obsolescent by 1891 but no doubt they had good reasons for doing so. The type had proved reliable in the first *Lord of the Isles*, fuel was not as dear as in later years, and initial cost may well have been less than that of a set of modern compound machinery. Similarly, the boilers were of the well-tried haystack type which had advantages of lightness, cheapness and rapid steaming, and these probably tilted the scales against the more sophisticated designs in vogue in the railway fleets.

The second *Lord of the Isles* was launched on Saturday, 25 April 1891 by Miss Mary Maclean, daughter of Mr William Maclean, of Plantation, chairman of the Glasgow & Inveraray Steamboat Company. On 20 May the vessel ran her trial trip to Inveraray, improving on her predecessor's best time by twenty minutes. A large party of invited guests travelled with her, some from Glasgow and others from Greenock and Gourock. During a speech after dinner, which was served on the return sailing, Mr John Henderson of the builders expressed his satisfaction at the vessel's having exceeded the speed of the earlier steamer; the only stipulation by the owners had been that she should better the performance of the first *Lord of the Isles* by half a knot per hour on the run from Rothesay to Inveraray, and this had been more than accomplished.

The new steamer's standard of finish was very high, as befitted one of the best tourist vessels on the Clyde. It must be remembered that social distinctions in the nineties were much more strongly marked than in the middle of the twentieth century and the *Lord of the Isles* was designed to cater for the rich tourist rather than the Glasgow working man and his family. Consequently, furnishing was on a lavish scale, as a contemporary description revealed :

The saloon is a large and tastefully decorated apartment, finished in white and gold. The furniture is of black walnut, upholstered in terra-cotta frieze velvet. For the convenience of passengers there are several writing tables. On either side of the entrance to the main saloon are ladies' and gentlemen's toilet rooms, and forward of the latter, on the starboard side, is the cloakroom. The lighting and ventilation of the dining room, which is situated on the lower deck, received the special attention of the builders. To give the place a cool appearance the floor is of polished hardwood. Gannaway's patent system of ventilation provides a constant supply of fresh air. The upholstery is of crimson morocco, and the furniture of black walnut. Handpainted floral designs adorn the side panels of the room. Elegantly designed revolving chairs are fitted at the tables. At the fore end on the starboard side is situated the pantry. . . . The galley is on the main deck above the pantry, and the dishes are sent down by means of an elevator. On the port side are the refreshment bar and bar-room. The accommodation for steerage passengers consists of a handsome saloon and dining cabin. The officers and crew are berthed forward under the main deck.

The machinery, while being of the same type as in the first *Lord of the Isles*, was nevertheless more powerful and incorporated some refinements. Steel paddle wheels were used, instead of wrought iron, although wooden floats were employed, and steel again took the place of wrought iron for the engine framing. A completely new feature was the use of Alley & McLellan's 'Sentinel' steam steering gear, placed on the engine room starting platform, and worked from the navigating bridge.

The new *Lord of the Isles* took up a strenuous daily round under the command of the popular Captain Donald Downie, who had commanded her predecessor for many years and had become a great favourite with the travelling public. Leaving Bridge Wharf, Glasgow, at 7.20 am, she called at various piers on her way down river, connecting with trains from St Enoch and Central stations at Prince's Pier and Gourock respectively, and then called at Kirn, Dunoon, Innellan, Rothesay, Colintraive, Tighnabruaich, Crarae, Strachur and Inveraray. Leaving again at 2.20 pm she returned to Glasgow after eight o'clock in the evening. After a few months it was found expedient to abandon the calls at Kirn and Innellan where delays often took place, but apart from these omissions this remained the daily round of the *Lord of the Isles* for years.

MacBrayne's veteran, the *Iona*, returned to Clyde service in June 1891 after a long spell in the hands of Hutson & Corbett. She had missed the season of 1890, but emerged rejuvenated after a very thorough reconstruction which involved the removal of her original horizontal boilers and their replacement by two large hay-stacks while other modifications involved the addition of a surface condenser. Being lighter than the old boilers, the haystacks allowed the *Iona* to float six inches higher in the water than previously. She was on trials on the measured mile at Skelmorlie on 6 June and achieved a speed of 17.1 knots. Resuming her station on 22 June, her daily duties during the summer consisted of the up early morning run from Ardrishaig, and the 1.30 pm down sailing from Glasgow, supplementing the mail service given by the *Columba* whose place she took at the beginning and end of the summer when traffic was lighter.

BUCHANAN CHANGES

As new ships came into service so too did old ones depart, and the same summer of 1891 which saw the arrival of the new *Lord of the Isles* saw the last of an old Buchanan favourite, the *Balmoral*, which was taken to Bowling harbour in April to be broken up. She was by then the veteran of the Clyde fleet, having been built as the *Lady Brisbane* as early as 1842 for service on the Largs and Millport station. As *Balmoral* she was well known on the Greenock, Helensburgh and Gareloch route until her withdrawal as the result of a serious mechanical failure. The other vessels in the Buchanan fleet, the *Guinevere*, *Vivid* and *Shandon*, appeared in the 1891 newspapers in happier circumstances when they made special sailings from Paisley on the occasion of the formal opening of the River Cart on 20 April after the completion of the deepening and widening operations which were intended

Page 107: The turbine era: *(above)* the first *Queen Alexandra* on the Campbeltown service; *(below)* the original propeller arrangement of the *King Edward.*

Page 108: Railway turbines: (above) the *Duchess of Argyll* as modified with bow plating in 1910; (below) the *Atalanta* passing Gourock.

to allow navigation as far as Paisley itself. The scheme cost some £130,000. It was the town's spring holiday, and many hundreds of citizens availed themselves of the special cruises, and it was recorded that hundreds more were left behind. The first two steamers sailed for Rothesay and intermediate piers, but the *Shandon* went to Garelochhead. All three were enthusiastically cheered by crowds of people lining the river banks. The work of making the river navigable had been considerable but in spite of the efforts of those responsible it has to be recorded that passenger sailings from Paisley never developed.

When the Glasgow & South Western Railway bought Captain Alexander Williamson's 'Turkish Fleet' in 1891 another old-established Clyde steamboat business came to an end, in this case only temporarily, for his son, John Williamson, went into business for himself within the year. Nevertheless the tradition of Williamson boats as such at Prince's Pier was broken, albeit they lingered in alien colours for a few years under the care of Alexander Williamson, jr. The four vessels comprising the fleet were old by the early nineties but all had been renovated in 1890. The little *Sultan* had been reboilered while the *Viceroy* and *Marquis of Bute* were equipped with surface condensers instead of the jet type while both, together with the *Sultana*, had been fitted with steam starting gear to improve manœuvring at piers. In the early part of 1891 the *Viceroy*, as we have seen, was subjected to even more radical alterations and improvements under the supervision of Alexander Williamson, jr, being rebuilt as a saloon steamer. Deck saloons were added fore and aft, giving a promenade deck 150 ft long. The original cabin aft was converted into a dining saloon seating sixty people and it was noted that 'the new deck saloon is decorated and upholstered in the most recherché manner, and a smoking room is fitted at the fore-end, while the ladies' cabin is aft'. Thus modernised, the *Viceroy* resumed her station, leaving Kames daily at 7.30 am for Glasgow, arriving at 11.45 am. She returned from the city at 1 pm, on weekdays to Kames and on Saturdays on a cruise to Loch Ridden. The *Sultana* and *Marquis of Bute* both operated from Prince's Pier in their last months as Williamson steamers, the former sailing to Glasgow and Port Bannatyne, and the latter to the Kyles of Bute and Ormidale. The *Sultan* was the auxiliary vessel, being used for luggage and cargo runs between Glasgow and Rothesay.

The Buchanan fleet was augmented in 1892 when a new paddle steamer entered service, replacing the *Guinevere*, which was sold for £4,400 to the Turkish government for service at Constantinople. Under the command of Captain Brock she left the Clyde on 17 February 1892 but never reached her destination. The second half of February was exceptionally stormy on the western seaboard of Europe and the *Guinevere* was overwhelmed in a gale in the Bay of Biscay and foundered, leaving no survivors. Captain James Williamson records in *The Clyde Passenger Steamer* (1904), that the disaster was witnessed by the crew of a Clan liner who were unable to render assistance. The new steamer which took her place was the last Clyde passenger vessel to be built by T. B. Seath & Co, of Rutherglen. She was launched on 14 May 1892 and named *Isle of Arran*. Her single diagonal, surface condensing machinery and haystack boiler were supplied by William King & Co. This was a smart steamer, designed for moderate speed and economy of working in sharp contrast to the railway boats of the same year. There were deck saloons fore and aft, to the full width of the hull, the navigating bridge was abaft the funnel, and the interests of safety at sea and passenger comfort were thriftily reconciled in the provision of a large awning over the promenade deck aft which, in the event of a mishap, was intended to serve as a life-raft. She entered service on 15 June 1892, sailing 8 am from Glasgow to Dunoon, Rothesay, Kilchattan Bay, and thence to the various places on the east coast of Arran—Corrie, Brodick, Lamlash, King's Cross and Whiting Bay. She returned from Whiting Bay at 1.50 pm, and followed the same route back to Glasgow.

By 1892 it was clear that the saloon steamer had come to stay and the North British Steam Packet Company sent the *Guy Mannering* to A. & J. Inglis, of Pointhouse, Partick, to be converted to the new style, leaving only the *Jeanie Deans* as a raised quarter deck vessel. The *Guy Mannering*'s quarter deck was cut away and a full deck saloon put in its place, described as having the sides finished in polished hardwood, with pilasters and capitals to match, while the ceiling was covered with Lincrusta Walton. There was a writing table, and the upholstery was 'richly finished'. What was discreetly referred to as 'the ladies' boudoir' was 'beautifully finished', with handpainted floral devices on the panels. There was no deck saloon on the fore-deck in the usually accepted sense, but a small, detached smoking cabin was provided. Having been reboilered in the previous year the *Guy Mannering* was expected to give a good account of herself but the alterations had the effect of changing the racer of old into something of a carthorse. Her speed was drastically impaired and the haste with which the North British company disposed of her was a sure indication that she was a disappointment in her new form.

RETURN OF THE VICTORIA

Undoubtedly the most interesting development in the private sector in 1892 was the reappearance on the Clyde of the former Wemyss Bay Company's flagship, the *Victoria*, under the aegis of the Scottish Excursion Steamer Company, Ltd. This vessel, built in 1886, was large and speedy, and it occurred to several people that there might well be room for her on the Clyde, operating excursions independently of any railway owners. Amongst these Hugh McIntyre, the Paisley shipbuilder, was prominent and appears to have taken a leading part in the formation of the new company. The *Victoria* was purchased on very favourable terms and sent to her original builders, Blackwood & Gordon, to be reboilered with larger haystacks and generally renovated. A special device, termed 'induced draught' was fitted in her funnels, probably to improve steaming, and a completely new electric light installation was provided. Thus improved, the *Victoria* made a trial trip on 28 April 1892 with a large party of guests. The prospects of the venture seemed as bright as the weather as the steamer sailed down the measured mile and, 'after a very satisfactory performance', cruised round Bute and the Cumbraes. The manager, Aird Wilson, said in an after dinner speech that the charters already arranged were sufficient to guarantee a dividend, and he proceeded to outline plans for cruises. The steamer was intended for Saturday afternoon excursions from Glasgow during May, after which daily cruises were to be undertaken from Greenock or Helensburgh to Ailsa Craig and Campbeltown, with occasional runs to Arrochar.

CAMPBELTOWN ROUTES

There had been a long history of indifferent service to the resorts in the west of Arran and Kintyre. The Campbeltown & Glasgow Steam Packet Joint Stock Company, Ltd operated a daily service from Glasgow to Campbeltown via Lochranza and Pirnmill with its screw steamers but it had established a virtual monopoly which, as we have seen, had been successfully defended against railway incursions in the parliamentary battle of 1891. Nevertheless, its service left something to be desired. There was much delay due to the handling of goods and livestock, as the following letter of complaint, which was printed in *The Glasgow Herald,* illustrates :

Glasgow, July 31, 1889.

Dear Sir,
It was my fortune this summer to spend my holiday at Lochranza, and until yesterday I found it in every way an enjoyable one. To avoid the overcrowding usual at the end of the month, I arranged to leave with the Campbeltown morning steamer, due at Lochranza, as per time bill, at 9 A.M. Accordingly I had my goods and chattels, wife and family on the pier at 8.45. Hour after hour passed and no word of any steamer. At twelve o'clock, however, the *Kintyre* put in an appearance, and such an appearance ! Not a square foot of space in the steerage for a passenger. Sheep were in front, behind these pigs, then bullocks, and finally herring boxes innumerable packed up to the top of the fore deck. Best, or, more accurately, worst of all, however, the after cabin was occupied by another flock of sheep, while the passengers had

to sit inside the small boats or on top of boxes. After shipping another couple of hundred herring boxes or so, and getting the passengers' luggage thrown in after us in a helter-skelter fashion, so as to block up entirely one side of the vessel, we set sail—the nine o'clock steamer—at one o'clock in the afternoon for Glasgow. As if this delay had not been sufficient, the *Kintyre* was put in to Dunoon, where a number of Volunteers were landed. Glasgow was reached at twenty minutes to eight, or nearly a round of the clock from the time the vessel was timed to leave Lochranza.

Now I think conduct of this description ought not to pass unnoticed. Complaints by passengers on this route are numerous. Will anyone be bold enough to say these are without foundation? If a steamboat company advertises to take passengers why should they be inconvenienced and put to serious loss by delay in this fashion? If there be a cabin in a steamer, why not let it be a cabin and not a sheepfold?

The other route to west Arran and Campbeltown was from Fairlie, in connection with the Glasgow & South Western Railway, direct to Lochranza and down Kilbrannan Sound. A screw steamer called the *Argyll* opened this route in 1885 but in 1891 the steamer being used was the *Herald*, an elderly paddle vessel which had been built in 1866. She was the subject of as much abuse as the Campbeltown company's fleet, the following letter which appeared in *The Glasgow Herald* being typical:

September 2, 1891

Sir,

. . . Passengers from the Fairlie route reached St Enoch's about 11 pm [on 31 August] nearly three hours late. Had this delay been exceptional it might have been overlooked, but it was simply the culmination of numerous delays during the month. If this admirable route is to be properly developed better arrangements must be made next season. A first-class, thoroughly equipped steamer should be put on, instead of a converted luggage boat with spliced hawsers snapping every other day, and endangering the safety of both passengers and crew.

It was evidently in response to public demand and possibly with the encouragement of the Glasgow & South Western Railway that the Scottish Excursion Steamer Company placed the *Victoria* on the route on 24 June 1892 to operate a regular daily service. She worked a complicated roster, evidently designed to free her for private charters during certain times of the day; unfortunately the constant adjustment of the timetable gave the vessel's owners a bad reputation. A correspondent wrote to the *Herald* to say that 'when rumours of an agreement between the Glasgow & South Western Railway Company and the *Victoria* reached the ears of old travellers, there was great jubilation over the prospect of at last having a smart paddle boat to convey us. How we have been disappointed. The steamer was worked in a manner that made it impossible for her to be a success on the Arran route.' A cause of much bitterness amongst holidaymakers on the west side of Arran was the abrupt curtailment of the service at the beginning of August. One infuriated traveller conducted a vitriolic correspondence with the superintendent of the Glasgow & South Western Railway complaining of breach of faith and threatening 'to adopt other methods of enforcing you to fulfil your contract with me', but the railway company replied blandly that the *Victoria* was not their steamer and that they could do nothing about the matter. After a chequered season, the *Victoria* ended the summer with a closing cruise round Cumbrae on 10 September and went into winter quarters.

She emerged on 1 May 1893 to commence her second season with a Bank Holiday excursion from Glasgow to Loch Ridden, and sailed on subsequent dates to the Kyles of Bute. From 29 June she was advertised to resume a full daily service from Prince's Pier and Fairlie, in connection with the South Western Railway, calling intermediately at Dunoon and Rothesay, and afterwards at Millport whence her sailings on Mondays, Wednesdays and Fridays were to Campbeltown, on Tuesdays Ailsa Craig, and on Thursdays Ayr and Culzean Bay. She returned to Fairlie and Greenock daily to connect with trains leaving for Glasgow at 6 pm and 7.15 pm respectively. Her Saturday duties consisted of an afternoon run from Glasgow to the Kyles. Destinations later in the summer were varied on Tuesdays and Thursdays to include Loch Fyne and Girvan, but the Mondays, Wednesdays and Fridays pattern of runs to Campbeltown remained, as did the Saturday afternoon trips. It was a less ambitious but much more sensible approach than in 1892 but there could be no gainsaying the fact that the Campbeltown route was poorly served and public meetings were held in the Arran villages of Lochranza and Pirnmill in July 1893 to protest against the lack of facilities. The *Victoria* was withdrawn at the end of the season and we hear no more of the Scottish Excursion Steamer Company, Ltd after that date. As no replacement was provided for the steamer, the Arran inhabitants were thrown wholly upon the Campbeltown & Glasgow Company's services in 1894.

The Glasgow & South Western Railway found it possible to dispose of a steamer in 1893 on the arrival of its new steamers *Minerva* and *Glen Rosa*. The *Sultan* and *Sultana* were offered for sale and on 14 March the company's minutes recorded the receipt of 'a letter dated 11th inst. from Captain John Williamson, Bridge Wharf, Glasgow, offering to purchase the Steamer *Sultan* for £500 . . . and authority given to offer the Steamer for £750 cash'. On 11 April, however, 'the General Manager reported that . . . he had offered this Steamer to Captain John Williamson at £750, which offer had not been accepted, and that acting on the instructions of Sir Renny Watson, with the approval of Mr Caird, he had accepted Capt. John Williamson's offer of £500, which was approved'. Thus the *Sultan* became a Williamson steamer again, reverting to her original colours, but bearing the new name *Ardmore*. The remainder of her Clyde career may be summarised at this stage. In 1894 she was bought by David MacBrayne for service on the Caledonian Canal, for which her size made her admirably suited, and under her third name, *Gairlochy*, she sailed on the canal until her accidental destruction by fire at Fort Augustus on Christmas Eve, 1919.

The spring of 1894 brought surprising developments in the privately-owned river fleet. *The Glasgow Herald* of 5 March carried a report that 'the popular Clyde steamer *Ivanhoe* is already getting into trim for the season, and will probably within the next week or two proceed to the Manchester Canal, on which she will make a trial of service for the two months or so prior to the opening of the Clyde season. This is in the way of an experiment on the part of Captain James Williamson, the result of which will determine future action.' The opening of the Manchester Ship Canal in 1894 brought a determined effort to inaugurate a passenger excursion service on the part of the Ship Canal Passenger Steamer Company (1893), Ltd, which chartered the *Ivanhoe* from the Frith of Clyde Steam Packet Company. As the report suggested, the *Ivanhoe* venture was experimental and she returned to the Clyde on 23 May to resume her Arran sailings. She was followed to Manchester in May by the two Buchanan steamers *Shandon*, which was renamed *Daniel Adamson* after the Ship Canal Company chairman, and the *Eagle*. Both of these vessels were bought from Buchanan, whose Clyde fleet was reduced thereby to only three steamers. The Manchester Canal passenger trade never developed successfully, and in 1895 the *Shandon*, still bearing her new name, returned to the Clyde, sailed for a short period, and was then withdrawn for breaking up. The *Eagle* alone of the three Clyde steamers remained in the south, and was broken up in 1899.

It will be recalled that the North British Steam Packet Company had had the *Guy Mannering* reconstructed with a full deck saloon in 1892. The *Jeanie Deans* was the last vessel in the fleet to retain the old arrangement of raised quarter deck, but in 1894 she too was given deck saloons fore and aft. Whatever the improvements in comfort, the same disastrous effect on her speed was noticeable as in the case of the *Guy Mannering*, and the *Jeanie Deans* no longer ranked as one of the greyhounds of the firth. The company had little alternative but to order new vessels to compete successfully with those of the Caledonian Steam Packet Company and the Glasgow & South Western Railway, a policy which, once decided upon, was pursued vigorously. The close of the 1894 season saw the *Guy Mannering* sold out of the fleet, being purchased by Captain Buchanan by whom she was given her third name, *Isle of Bute*. Further reconstruction was

carried out and she emerged for the 1895 season with a full deck
saloon forward instead of the detached smokeroom which had
formerly occupied the foredeck.

THE BROOMIELAW AT THE FAIR

Despite the inroads of the railway fleets and the often noisome
condition of the upper Clyde, traffic from the Broomielaw could
still be remarkably heavy and at the Glasgow Fair Holiday of
1894 there was an impressive array of steamers. It was recorded
that on Fair Friday, 13 July, eleven steamers had amongst them
carried about 7,000 people but on Fair Saturday thirteen vessels
sailed from the city to coast resorts. Even this formidable fleet
was inadequate for the traffic. It was recorded that the steamers
arriving at Greenock in the early morning carried enormous
crowds, and that the *Lord of the Isles* and *Columba* had to turn
away most of the passengers waiting to join at Prince's Pier. The
former, indeed, had to discharge some Dunoon passengers and
their luggage at Gourock to allow a few more Inveraray passengers on board. The Campbeltown company's *Davaar* had to
leave a hundred people at Gourock, but they were picked up later
by the *Kintyre*.

Scenes of this kind at the Broomielaw were becoming less
common by the middle nineties as the popularity of the new
railway steamers grew, and as the years went on the importance of the up river route steadily declined. Time was to show
that summer excursions could still be profitably operated by
owners such as the Buchanans, but the responsibility of working the bulk of the winter traffic eventually fell to the railways
and scenes such as the Fair Holiday rush of 1894 recalled the heyday of the Broomielaw route in the seventies and eighties rather
than reflected its future potential.

CHAPTER FIVE

VICTORIAN CLIMAX
(1895-1900)

THE later nineties were the years of the fiercest competition for Clyde passenger traffic. From 1895 to 1899 new steamers were added annually to the various fleets. It was the high noon of paddle steamer design, and many notable railway ships entered service. Amongst privately-owned steamers, the *Victoria* brought upon herself the wrath of Sabbatarians, while the *Davaar* almost ended her career in strange waters. The North British Company was fortunate in narrowly avoiding tragedy when one of its fleet was wrecked off Arran, but as the century ended these spectacular events contrasted sharply with a sudden lull in activity in 1900.

CRAIGENDORAN ADDITIONS

There had been many notable changes in the Clyde coast services since the end of the eighties, but in 1895 competition amongst the steamer operators entered an even more uninhibited stage. During the years of Caledonian and Glasgow & South Western expansion the North British Steam Packet Company stood rather aloof from the main context; possibly Robert Darling, the general manager, hoped to consolidate his company's position without recourse to unnecessary expense in building new steamers. The *Jeanie Deans* and *Guy Mannering* were still relatively new vessels, recently modernised, and he may well have considered them perfectly adequate to maintain the Rothesay services. The *Lady Rowena*, even newer, was primarily an excursion steamer on which no great demands of speed were normally made, while the *Lucy Ashton*, *Lady Clare* and *Diana Vernon*, although small, were saloon vessels well able to match their rivals on short distance routes. On the face of it, the North British was dealing with its problems rather more sensibly than its south-bank competitors, and at much lower cost. But standards had risen faster than perhaps was appreciated at Craigendoran, and an impatient public clamoured for steamers which were not only more comfortable but also faster than their predecessors. In recognising the demand for saloons by modifying the *Guy Mannering* and *Jeanie Deans*, the racing capacity of those redoubtable vessels had been seriously impaired and they were no longer capable of their original speed. There can be little doubt that Craigendoran traffic began to fall, forcing the North British Railway and its Steam Packet Company into a policy of new construction which was pursued with great thoroughness until the close of the century.

The first step was the disposal of the *Guy Mannering* to Captain W. Buchanan in 1894. She was renamed *Isle of Bute* and went up river as consort to the *Isle of Arran* on services which required no great speed. Unlike railway steamers, the Broomielaw boats simply *had* to pay their way, and in this respect the *Isle of Bute* must have been a sound investment, for she remained in the Buchanan fleet for eighteen years, generally sailing to Rothesay and on short cruises from that port. Before entering on her new duties she was further improved by the provision of a larger forward saloon, and the conversion of the lower saloon aft into dining accommodation.

To replace her, the North British Company ordered from Barclay, Curle & Co a saloon steamer similar to the modernised *Jeanie Deans*. The new vessel was larger and faster than the *Jeanie* although the conservatism of the management was reflected in the fact that the machinery was of the well-tried single diagonal type and the boiler once again of low pressure, haystack design. The new steamer was launched on 4 April 1895 and named *Redgauntlet*. Although generally similar in style to the other North British steamers, she was nevertheless readily identifiable by the rather large observation windows in the saloons. The foresaloon, for the use of steerage passengers, was narrow, with alleyways round it. Under the command of Captain Dan McKinlay, the *Redgauntlet* was placed on the Craigendoran–Rothesay route with the *Jeanie Deans* as her consort.

A second new steamer appeared in the same year. This was the *Dandie Dinmont*, the second of that name at Craigendoran. She was built by A. & J. Inglis, of Pointhouse, Partick, and was launched on 10 May 1895, being described as 'a beautifully modelled paddle steamer . . . built to the specification of Mr Robert Darling, the company's manager'. The *Dandie Dinmont* was a smaller version of the *Redgauntlet*, with similar boiler and machinery and deck saloons fore and aft. She took up the sailings to Kilcreggan, Hunter's Quay and Holy Loch piers under the command of the popular Captain D. McNeill and her arrival allowed a redistribution of duties within the North British fleet. The *Redgauntlet* and *Jeanie Deans* maintained the express services to Rothesay; the *Lady Rowena* continued to sail to Arrochar in connection with Loch Lomond services to Tarbert; the *Dandie Dinmont* was the new Holy Loch steamer; and the two small steamers *Diana Vernon* and *Lady Clare* operated respectively on the Gareloch station and on a new ferry service between Greenock (Prince's Pier) and Craigendoran, the latter intended to afford a link between Renfrewshire and Ayrshire towns and the newly opened West Highland Railway to Fort William. The *Lucy Ashton* was freed to act as standby steamer and ran various excursions and charters.

CLYDEBANK STEAMERS FOR 'CALEY' AND 'SOU'WEST'

The first addition to the Caledonian fleet for several seasons also appeared in 1895. During 1894 the directors had discussed the desirability of ordering a new boat, and Captain James Williamson was instructed to prepare specifications and estimates. Having submitted these to various Clyde shipbuilders, their tenders were duly considered at the end of November. Amongst them was one from J. & G. Thomson, Clydebank, and the company's minutes recorded that 'the offer of the latter being the lowest, Tender was accepted—namely £17,500 and Secretary instructed to accept'. At this stage there was a new development; it appears that the shipbuilders suggested that they might build a larger steamer, differing from the specifications, for on 20 December 1894 it was minuted that:

> The position of the new Steamer ordered from J. & G. Thomson, Limited was considered and the Secretary instructed to accept the alternative offer made by that company for a larger steamer for the sum of Nineteen Thousand, nine hundred pounds (£19,900) and which is to be named the 'Duchess of Rothesay', Lord Breadalbane agreeing to get the sanction of HRH The Princess of Wales for our doing so.

The name, of course, was the principal Scottish title of the consort of the heir to the throne and the new Caledonian steamer took her name from the future Queen Alexandra.

J. & G. Thomson had not previously built any steamers for the company, but their Glasgow & South Western vessels had been very successful. In the case of the *Duchess of Rothesay* they excelled themselves, producing one of the most attractive and generally useful steamers ever to sail on the Clyde. It has to be admitted that although the early steamers acquired by or built for the Caledonian company were successful, with the exception of the luckless *Meg Merrilies*, none could really be regarded as beautiful. The *Duchess of Hamilton* was the best in this sense, but her lack of sheer made her look stiff, while the funnel was insufficiently raked to enhance her appearance. But in the *Duchess of Rothesay* everything came right, and the result was a joy to look upon. J. & G. Thomson modelled her on the *Slieve Bearnagh*, a steamer which they had built in 1894 for the Belfast & County Down Railway's excursion services between Belfast and Rathlin Island or Killough. She was of uncomplicated design, propelled by compound diagonal machinery and fitted with a double-ended boiler. All of these features were repeated in the Caledonian steamer which was of similar dimensions. After a long flirtation with Navy boilers the double-ended type represented something of a new departure, but it was simply equivalent to two of the

Navy type. A straightforward two cylinder compound engine was used, probably to avoid any disappointment in speed such as had occurred with the triple expansion *Marchioness of Lorne* in 1891.

Work was pushed ahead rapidly, so that payment of the first instalment of the contract price was authorised by the Caledonian directors on 5 February 1895, the second on 26 March and the third on 7 May. In the meantime the steamer had been launched on Saturday, 20 April, and so rapidly was she completed that her trials were run on Friday, 17 May when she achieved a mean speed of 18 knots between the lights. In general, she was in much the same style as the *Duchess of Hamilton*, with promenade deck carried forward to the bow although, as in the case of the older boat, and unlike her Irish half-sister, the plating was not carried above the mainrail, leaving the bow open for the handling of mooring lines at piers from main deck level. The *Duchess of Hamilton*'s lines had been spoiled to some extent by insufficient sheer, but the *Duchess of Rothesay* gained immeasurably in this respect. Her funnel was well proportioned and raked, and the bridge was placed much further forward than that of the older *Duchess*. The result was quite unusually attractive, and it was generally acknowledged that aesthetically the *Duchess of Rothesay* was one of the finest steamers on the river. A new and very elegant design of paddlebox was incorporated, rather similar to those of the South Western steamers, although the Caledonian practice of fitting specially carved and finely painted armorial bearings was continued.

The press gave the usual full description of the new steamer, noting, *inter alia*, that 'an improvement on previous ships of the company has been effected in the erection of light teak deck-houses at the head of each stairway leading from the promenade deck, so as to prevent rain and spray being blown down in stormy weather. The first class saloon is a very handsome apartment, panelled with mahogany and satinwood and lighted by large square windows, the tops of which are hinged to give thorough ventilation. . . . The vessel is fitted with the latest improvements, including Kerr and Rayner's feed-heater, Alley & McLellan's patent "Sentinel" horizontal steam steering gear and feed water-filter, warping capstans forward and aft, electric installation throughout, and docking telegraphs. . . .' The *Duchess of Rothesay*, commanded by Captain McPhedran, formerly of the *Marchioness of Breadalbane*, entered service as additional steamer on the Ardrossan and Arran route for a short period at the end of May, 1895 and on 1 June took up regular duties on the Rothesay station, sailing in alternate weeks from Gourock and Wemyss Bay.

At first she had little opportunity of showing her mettle, for in early May the Caledonian and Glasgow & South Western companies made arrangements for avoiding unnecessary competition, involving retiming of the principal expresses from Glasgow, slight extensions of overall journey times, and interchangeability of season tickets. These moderate reforms were met by a storm of protest from the public. 'I beg to draw your attention to the absurd arrangement come to between the Caledonian and Glasgow and South Western Railway Companies, whereby the travelling public to the coast are the sufferers,' wrote a correspondent in *The Glasgow Herald*. 'Although the Caledonian Company can with ease convey its passengers to here [Dunoon] from Glasgow in 50 minutes we are obliged under this new arrangement to occupy an hour. The disappointment of the season-ticket holders is great, and how the Caledonian ever came to make such a soft agreement is a mystery, after being at the expense of building such a fine swift boat as the *Duchess of Rothesay*. The complaints are universal. . . .' His indignation was shared by another traveller who wrote : 'I can endorse what is said as to the dissatisfaction felt by the daily travellers, for not one have I heard but is astonished at the stupidity of it. Why should the Caledonian take a back seat when by new boats and good running they had got the ball at their foot?'

Despite the inter-company agreement, however, the racing spirit was by no means extinguished, and the *Duchess of Rothesay* found herself involved in a joust with no less redoubtable an opponent than the *Glen Sannox* on 27 June. A passenger on board the *Rothesay* recorded the circumstances :

> The Caledonian steamer *Duchess of Rothesay* was on an excursion trip round the Island of Arran. Her last call for passengers was at Millport, from which she was making her way across the channel for the Garroch Head. When she was about half-way from Cumbrae the South-Western steamer *Glen Sannox* was seen evidently lying in wait, and as the Duchess came near the former made a circuit and came up on the rear of the *Duchess* on the south side. From this point to the north end of Arran, near Loch Ranza, there was a keenly contested race. On board the *Duchess* every pound of steam was forced into action, with the result that the hull of the vessel quivered from stem to stern because of the tremendous pressure on the machinery. Excess steam was blown off with a deafening noise, and the passengers, who by the way were railed off from the wings of the steamer by ropes, and who had come out to enjoy a pleasant excursion sail, were compelled for half an hour to endure an experience of excitement which I can best describe as like what sitting on the edge of a volcanic crater would be that was every moment liable to have an eruption. . . . If there is no law to prevent passenger carrying steamers from indulging in such high jinks, I hope that the directors of the respective companies will instruct its suppression.

He wrote in vain.

MORE NEW VESSELS

By 1895 Captain John Williamson's Broomielaw trade was sufficiently promising for him to order the first steamer on his own account, all of his existing tonnage being bought in from other owners. The new ship was named *Glenmore* at her launch from Russell & Co's yard, Port Glasgow on 9 April. She was a plain-looking steamer, of moderate size, propelled by compound diagonal machinery. The promenade deck was carried forward to the bow and the topside plating carried up to that level at the bow itself, but a large section was left open further aft, immediately forward of the boiler. It is understood that the bow required to be

Page 125 : Craigendoran elegance : the *Marmion*—the final development of the North British paddle steamer.

Page 126: Broomielaw favourites: *(above)* John Williamson's *Strathmore* leaving Dunoon; *(below)* Buchanan's *Eagle III* going 'doon the watter'.

plated up in view of the steamer's intended use on the exposed Campbeltown run, but the long gap in the plating was a strange feature that was only once repeated on the Clyde, and that in the same season. Certainly it gave the *Glenmore* a strange appearance. Her early work, in addition to direct sailings from Glasgow to Rothesay, included the afternoon trips to Kyles of Bute piers as far as Auchenlochan and Kames and Saturday cruises round Bute, while at Glasgow Fair holidays she was on special sailings from Glasgow to Lamlash via Dunoon, Rothesay, Largs, Millport and Brodick in opposition to Buchanan's *Isle of Arran*.

In response to mounting public demand for a good steamer service to the west side of Arran, an entirely new venture started in 1895 when the *Culzean Castle* was placed on that route. It will be recalled that the Scottish Excursion Steamer Company's *Victoria* had made calls in 1892 and 1893, but apart from the Campbeltown Company's services nothing more had been done in 1894. The *Culzean Castle* was a very rare example of an English-built steamer bought second-hand for Clyde excursion service. A product of the Southampton Shipbuilding & Engineering Co, she had been built in 1891 for the Bournemouth, Swanage and Poole Steam Packet Co, but was not regarded as being suitable for their services. Her original name of *Windsor Castle* was altered when she came to the Clyde. A new concern, the Glasgow, Ayrshire & Campbeltown Steamboat Coy, Ltd was formed to operate her, under the managership of D. T. Clark. The *Culzean Castle* was a large, handsome paddle steamer—very nearly as long as the *Duchess of Hamilton*—and was driven by triple crank, triple expansion machinery, the first of its type in a Clyde steamer. As first built, she had a raised forecastle with a well deck arrangement forward of the funnel but in due course the promenade deck was carried forward to the bow leaving the sides unplated in the same manner as the *Glenmore*. The *Culzean Castle* was advertised to enter service on 1 June 1895. Her regular daily run involved a 9 am departure from Prince's Pier, whence her route lay via Dunoon, Largs, Fairlie and Keppel (Millport) to Lochranza, Pirnmill, Machrie Bay and Campbeltown, arriving there at 12.30 pm. Special excursions were run in connection—the Glen Sannox Coach Tour allowed tourists to go ashore at Lochranza for luncheon in the new hotel before driving across the remarkable scenic road to Sannox and Corrie, returning in time to join the steamer on her return sailing. Connections by road to Shiskine and Blackwaterfoot were available to passengers landing at Machrie Bay, while at Campbeltown tourists could travel by coach to Machrihanish Bay on the Atlantic shore of the Kintyre peninsula. It will readily be recognised that the new venture with the *Culzean Castle* gave the Glasgow & South Western Railway what it had been denied by Parliament, namely, a through service to Campbeltown and west Arran resorts. The morning departures from Prince's Pier and Fairlie were in connection with the 7.45 and 9 o'clock trains from St Enoch, and the return service connected with expresses to Glasgow. Unfortunately for the company's image, the *Culzean Castle* was disabled at a most awkward time in the summer of 1895, suffering a machinery breakdown at Keppel pier on her outward run on Glasgow Fair Saturday. She was quite unable to proceed to Campbeltown and ultimately had to be towed up river for repairs. Her passengers, after spending much of the day in Millport, were eventually taken to Fairlie and thence to Campbeltown by the hastily chartered *Isle of Arran*. It was one of a series of breakdowns which finally told against an otherwise splendid steamer, and the *Culzean Castle*'s career on the Campbeltown route was brief.

Undoubtedly the most dramatic mishap of 1895, however, involved the *Davaar*, of the Campbeltown & Glasgow Company, which, on Friday, 7 June, took a party of some five hundred excursionists from Campbeltown to Belfast. She left in clear weather but ran into thick sea mist shortly after passing the Mull of Kintyre and about half past ten ran aground on Brigg's Reef, in Belfast Lough, in visibility estimated to be as little as twenty yards. Although hard and fast on the rocks the steamer was not badly damaged and the sea was completely calm. Nevertheless there was panic on board the *Davaar*; a passenger afterwards spoke of 'women and children crying and screaming one upon the other and each and all displayed the greatest possible signs of fear'. Order was restored when it was seen that there was no immediate danger and when the fog cleared shortly afterwards the passengers were quickly ferried ashore by local fishermen and coastguards. The passage home proved more of an ordeal than the stranding. It was hoped that the paddle steamer *Slieve Donard*, of the Belfast & County Down Railway (cf. Ch. III), might be chartered to take the passengers back but the Belfast Board of Trade would not grant a provisional passenger certificate for the open sea voyage and in the event the excursionists had to return in Messrs Burns' steamer *Dromedary*, not to Campbeltown, but to Greenock. Here they were disembarked at four o'clock on the following morning at Prince's Pier where waiting rooms were placed at their disposal until they were taken on board the Campbeltown Company's *Kintyre* nearly five hours later. The tired and hungry party, consisting mainly of 'the better class working people' (*The Glasgow Herald*) overran Greenock's coffee shops as soon as they opened. Some of the party were doubly unlucky; having lost their original excursion tickets in the panic after the *Davaar* went ashore, they were required to pay their fares over again in the *Dromedary* and the *Kintyre*!

The unfortunate *Davaar* was eventually towed off the rocks and repaired in Belfast before returning to the Clyde. She had been lucky, for many vessels had come to grief on Brigg's Reef. One such was the *Emily*, sunk in 1882, and part of her wreckage still constituted a hazard. While the Belfast tug *Ranger* was engaged in trying to tow the *Davaar* off the reef, she struck the *Emily*'s sternpost and was holed and sank in half a minute.

When the Glasgow & South Western Railway steamer *Neptune* was placed on the new service from Prince's Pier via Rothesay and the Kyles of Bute to Corrie, Brodick and Whiting Bay in 1894 she evidently gave satisfaction and led the directors to consider building a new steamer for that important tourist route. On 10 September 1895 it was minuted as follows:

> The General Manager reported that in order to deal successfully with the Coast Traffic it would be necessary to provide a new

Steamer and it was resolved to negotiate with Messrs. J. & G. Thomson for a suitable steamer at a cost of from £22,000 to £23,000.

On 19 November a further minute recorded 'that a new Steamer had been contracted for with Messrs J. & G. Thomson, for £23,550, for delivery 1st May next, which was approved', and on 11 February 1896 the directors 'resolved to call the new Steamer "Jupiter".' She was launched on 21 March 1896 and ran her trials on 16 May under the supervision of Robert Morton. The average of two runs over the measured mile at Skelmorlie was 18½ knots, while a four hours' steaming trial was completed at an average speed of just over 18 knots. The *Jupiter* was one of J. & G. Thomson's happiest productions for the Clyde service The *Glen Sannox* excepted, she was the largest and most power- ful vessel yet to appear in the South Western fleet. Like the *Glen Sannox*, she was fully plated forward with the promenade deck carried to the bow, but she had only one funnel. One feature to which the company steadfastly clung was the placing of the bridge abaft the funnel, but this did not impair her good looks. Her machinery was of compound diagonal type, steam coming from a double-ended boiler. Mechanically, therefore, the *Jupiter* was a more powerful, South Western version of the *Duchess of Rothesay* and *Slieve Bearnagh*. She was placed on the Kyles of Bute and Arran service and rapidly established herself as a general favourite.

Her arrival allowed the *Sultana* to be disposed of, as she was by now completely obsolescent. At first it seemed that she was destined for the south of England; a minute of 7 April 1896 recorded that:

> . . . the sale of this vessel had been advertised . . . and it was remitted to . . . the General Manager to deal with the offer made by Mr Thomas Martin, Hoe Field House, Plymouth, for the purchase of the vessel at the upset price of £1,500.

However, there was no further reference to this transaction, and the next minute dealing with the *Sultana*, dated 11 August

1896, simply 'reported that this Steamer had been sold to Captn. John Williamson, Glasgow, for £750'.

In spite of being just over twelve years old, the North British steamer *Jeanie Deans* was no longer abreast of the times by the end of 1895, even after the expensive alterations of the previous year. She could not maintain her 17½ knots which, by the mid nineties, was inadequate for a top-class steamer, and the decision was taken, no doubt with regret, to replace her. She was therefore sold to the Derry & Moville Steam Packet Company, Ltd, of Londonderry, and spent two seasons in Irish ownership. Nevertheless, as we shall see, her Clyde career was far from finished.

As a replacement, the North British Company ordered an improved version of the *Redgauntlet* from A. & J. Inglis, of Pointhouse. The new steamer was launched on 30 March 1896 and received the name *Talisman*. She was the same length as her immediate predecessor, but slightly broader, and was similar in appearance to the *Redgauntlet* although that vessel's large observation windows were not repeated in the new ship. Both had the navigating bridge between the paddleboxes, and there were alleyways round the fore saloons, standard features in the North British fleet at this period. The *Talisman*'s saloons were described as being 'richly and chastely finished in pleasing colours both as regards the decorations and furnishings'. She went on trials late in May 1896 and was credited with a mean speed of 18.7 knots over the Skelmorlie mile which, for a steamer of her type—she was driven by the now old-fashioned combination of single engine and haystack boiler—was very satisfactory to builders and owners. During a cruise with a party of officials and guests on 26 May John Inglis, of the builders, presided at dinner and referred in a speech to the difficulty of satisfying at the same time the travelling public's insistence upon higher speeds and greater comfort and the owners' desire for economy. 'He thought it probable that the *Talisman* was the fastest steamer of her size ever built at the price, and to obtain the results they had witnessed was a considerable tax on the ingenuity of the builders.' Unfortunately, no records are available of the cost of the *Talisman* and *Redgauntlet*, but those who knew these ships in their early days are unanimous that they must have cost much less to build than their Caledonian and Glasgow & South Western contemporaries. The North British Steam Packet Company was quite obviously determined to avoid extravagance as far as possible. The contrast between its outlook and that of the South Western company is underlined by the examples of the *Talisman* and *Jupiter*. Built in the same year, each gave long service to her owners—thirty-nine and forty years respectively—and appeared to be equally satisfactory. Despite the difference in first costs, however, the speeds were for all purposes the same, and the North British directors could legitimately congratulate themselves on the acquisition of a very good steamer at an economical price.

The *Talisman* entered regular service on the Rothesay route on 1 June 1896, commanded by Captain McKinlay, and with two fast vessels available the North British was at last able to compete effectively with its rivals at Gourock and Greenock. Nevertheless, the Craigendoran route was at a disadvantage in respect of the railway portion of the journey from Glasgow, a source of delay to all coast trains being the single line tunnel immediately west of Dalreoch, on the Helensburgh line. This was worked on the train tablet system, and all trains had perforce to slacken speed to pick up and drop the tablet at each end of the tunnel. This caused much loss of time. Powers had been taken to double this section as early as 1877, but it was not until the nineties that the traffic increased to such an extent as to make this necessary. Work began in 1896 and involved the opening up of a part of the existing tunnel at each end and construction of a second single line tunnel to the south of it. On completion of the work, up trains used the old tunnel and down traffic the new one. The work was done in solid rock throughout and cost about £30,000 to complete, but it was possible to accelerate the boat trains and avoid delays after it was brought into use on 16 May 1898.

Meanwhile, another important improvement had taken place

on 13 May 1897 when a new cut-off route to the coast was opened to passenger traffic. It was a short section connecting the existing Clydebank branch, from a junction just east of the terminus, with the Helensburgh line at Dalmuir and on completion of the new line the North British had an alternative, shorter route to the coast, allowing still further acceleration of the Craigendoran traffic.

There was a lull in construction in 1897 as far as the railway companies were concerned, but the Caledonian fleet was augmented in May by the purchase of the *Ivanhoe*. Captain James Williamson had retained his connection with the Frith of Clyde owners when he was appointed Secretary of the Caledonian Steam Packet Company, and to all intents and purposes the *Ivanhoe* had operated as a unit of the Caledonian fleet for several years. The arrangement might have continued indefinitely but for the opposition of the *Jupiter* on the Kyles of Bute and Arran run. Not only was she more modern and faster than the *Ivanhoe*, but in the eyes of many members of the public she possessed the considerable advantage of a licence to sell alcoholic liquor. The *Ivanhoe* suffered in the contest and brought about the decision of the Caledonian directors to take over the route themselves. Thus, a minute of 23 March 1897 recorded a 'Report . . . in regard to purchase of the "Ivanhoe", and it was resolved to purchase that steamer at the price of £9,000, including furnishings in Steward Department, and all her belongings. Captain Williamson undertaking not to own or run any other steamer, or do any other business, but in connection with the Caledonian Steam Packet Company.' The final part of the minute recalled an earlier one of 25 February 1896 concerning Captain Williamson : 'The salary of the Secretary and Superintendent was considered, and it was resolved that, as from 1st January last, the salary shall be at the rate of £750 per annum, in respect of which he shall give his whole time to the Company's service as soon as his connection with the "Ivanhoe" is terminated.'

The Frith of Clyde Steam Packet Company was wound up, and so finished a courageous attempt to operate Clyde steamers on temperance principles. In the Caledonian fleet there was no room for a vessel which differed radically from the others in this respect, and the *Ivanhoe* was duly licensed. Her place on the Arran route was taken immediately by the *Duchess of Rothesay*, which was better able to compete with the *Jupiter*, but Captain Williamson himself recorded at a later date that the combined traffic of the two railway-owned steamers never equalled that of the *Ivanhoe* in her heyday. The public and press were inclined to regard the *Ivanhoe* experiment as a failure, but in looking back after many years it can be seen that the syndicate of 1880 did much to raise the standards of Clyde steamer operation. The ultimate failure of the 'teetotal' venture disguised the valuable achievements of the Frith of Clyde Company, amongst which may be numbered an improvement in behaviour on steamers, and the setting of high standards of comfort and service, all of which led to a general improvement on the firth.

The main events of 1897 concerned private owners. Captain John Williamson's trade continued to expand and he again ordered new tonnage, going back to Russell & Co, of Port Glasgow, for two identical vessels, to be named *Strathmore* and *Kylemore*. In the meantime, at the end of 1896, he had sold the new *Glenmore*, which had been in his fleet for only two seasons, to a firm in Siberia. It is improbable that the steamer had been unsatisfactory, and it may reasonably be concluded that he had simply taken advantage of a rather lucrative offer for the ship. Strangely enough, before the two new steamers of 1897 were completed, the second (*Kylemore*) was sold to a south coast concern, the Hastings, St Leonards and Eastbourne Steamboat Company, Ltd, for £13,750, and was renamed *Britannia*, and once again it may have been that John Williamson was able to get a good bargain. Whatever the true facts, the net result was that the *Strathmore* replaced the *Glenmore* and ran in 1897 with the *Benmore* and *Sultana* as her consorts. The *Strathmore* is noteworthy in having been the last new Williamson steamer to carry the old funnel colours, black with a white band, used by Captain Alexander Williamson, sr, from 1862, and continued by his son, for in June 1898 John Williamson changed his colours to all-white, with black top, the livery of the old Wemyss Bay steamers.

SUNDAY BREAKING

Undoubtedly the most celebrated steamer of the 1897 season was the old favourite of that fleet, the *Victoria*, which reappeared on the firth, this time in the ownership of a new concern, The Clyde Steamers, Ltd. Within a short time following her entry to excursion service from the Broomielaw the vessel was notorious as 'the Sunday breaker'. For some fifteen years no passenger vessel had plied on the river on Sundays, following unhappy experiences in mid-Victorian years, when the main motive activating Sunday operators was to sell liquor to those denied it ashore. There had been understandable indignation in coast towns when 'herds of savages and drunken rabbles' descended upon them to shatter the peace of summer Sundays and eventually Sunday boats were withdrawn. Nevertheless the unhappy memories lingered, and found their expression in the famous by-law no 5 of the Dunoon Commissioners, submitted to the Board of Trade for its approval early in 1897 :

> No steamer or other vessel shall be permitted to land or embark passengers at the pier between 12 midnight on Saturdays and 12 midnight on Sundays without the special sanction of the Commissioners, under a penalty by the party or parties in charge of said steamer or other vessel of a sum not exceeding £5 for each passenger landed or embarked in contravention thereof.

Such a ruling was not to the liking of Andrew Dawson Reid, the manager of The Clyde Steamers Ltd. This energetic and resourceful character became well known on the Clyde for a number of years at the turn of the century, operating several steamers in succession on excursion work from the Broomielaw with a panache and flair rivalling that of James Williamson himself. Reid and his associates held the view that Sunday excursions could be run without the indecorous scenes of earlier years, and they wrote to the Dunoon Commissioners to object to the legality of the proposed by-law, asking for the pier to be opened to receive passengers from the *Victoria* on Sundays. They met with a point-blank refusal. Nothing daunted, the steamer proprietors announced that their vessel would call at Dunoon on her opening cruise on 9 May 1897.

The *Victoria* left the Broomielaw at 10 o'clock amidst scenes which were to characterise her sailings throughout the whole summer. In fine weather six hundred passengers—mostly young men, with 'a fair sprinkling of ladies'—set off for a cruise round the Cumbraes, watched by large crowds all the way down river. Many people had gathered to watch her arrival at Dunoon, but there were no pier hands to take the ropes and her skipper, John McLachlan, made no attempt to get alongside in view of a strong wind blowing up firth. On the return run, however, the steamer was able to call and over twenty passengers were landed. The gates were locked and it was not for a considerable time that they were released. Following the incident, it was announced that the Board of Trade would hold an official inquiry into the new by-laws. The steamer owners felt confident of their position and noted with satisfaction that their policy of selling no intoxicants on board on Sundays would prevent any abuse of their service. On the following Sunday the gates were still locked against passengers, who had to satisfy themselves with lodging a formal protest with the local superintendent of police who, with his constables, had been called out to prevent any disturbance.

At a meeting of the Dunoon Commissioners held on 17 May it was apparent that a minority held the opinion that to close the pier was wrong, either on a point of principle or as a practical measure, and there were some lively exchanges on the subject of a petition said to have been signed by 1,701 ratepayers in support of the closure, Commissioner Crosbie in particular questioning whether all of the signatories were *bona fide* ratepayers. A request from the steamer company that a counter-petition be allowed to lie at the pier was peremptorily dismissed, and the meeting closed with the reading of a letter stating the owners' intention to call again on the following Sunday. There were no further incidents of any moment, however, until after the official inquiry at the end of May and, indeed, throughout June, when passengers were landed at Dunoon by locally hired boats. But the Glasgow Fair brought a new flare-up; the *Victoria's* owners had announced their intention of landing passengers at Dunoon pier on Sunday, 18 July. Excitement in the burgh was intense, and as early as eleven o'clock crowds began to assemble at the pier gates until, by a quarter to one, it was estimated that as many as 12,000 people had come to watch the proceedings. About a hundred passengers left the steamer on her arrival, led by Graeme Hunter, of the *Victoria* company, who stated his intention of removing the barrier at the gates, and assumed full responsibility. A determined attempt was then made to burst open the gates. The crowd in Dunoon, in full sympathy with the imprisoned travellers, attempted to close in to assist them, but were pushed back by the police, seven or eight in number, who were in attendance. Their action merely irritated the crowd and the police were hustled about until they drew batons and threatened to use them. At this critical moment the gate was broken open and the travellers were able to enter the town, whereupon the impending riot broke up. In the evening no difficulty was experienced and the *Victoria* took on board about 150 people.

The decision of the Board of Trade was given during the following week, and turned out to be a masterpiece of diplomacy. While passing the new by-laws, including the fifth, the Board made no comment on the legality of the attempt to prevent Sunday calls! Undeterred, Dawson Reid and his colleagues wrote to the Commissioners to say that they intended that the *Victoria* should continue to call and that the Dunoon authorities should take steps to test the matter in the courts. They were supported in a leader in *The Glasgow Herald* which counselled moderation and suggested that the company should collaborate with the Commissioners by landing a single passenger, in respect of whose landing a test case could be brought against the company so that the whole subject could be legally thrashed out.

Unfortunately the majority of the Commissioners were determined to insist upon Sunday closure, come what may, and when the *Victoria* approached Dunoon on 25 July another large crowd awaited her arrival with keen anticipation, the numbers again being estimated at about 12,000. The *Victoria*'s owners were anxious to avoid any unseemly conduct, and when she berthed at the pier three passengers, Messrs Austin, Hubner and Angeletti, made their way to the entrance gates and demanded admission to the town respectively 'as an Englishman, as a Scotsman, and as a foreigner'. Having made their dignified protest the trio returned to the steamer but at this stage a crowd rushed down the pier and forced their way into Dunoon. On the steamer's return run the crowds smashed the pier gates and a clash with the police seemed probable, but this was avoided after the *Victoria* took her passengers aboard and sailed off, amidst loud cheers.

Additional police were called into Dunoon on 1 August to prevent any serious riot, but the *Victoria* disembarked passengers at Kirn with her own lifeboats and no attempt was made to use Dunoon pier. At this time *The Glasgow Herald* printed a letter suggesting that religious services should be held on board the *Victoria* to save the lost souls who sailed on Sundays. The Rev Alfred Hargreave, an English clergyman, rashly offered to conduct such a service—'Many a Sabbath-breaker might be thus brought back to the fold . . . I think the boat might be used as a great means of grace to those who probably never darken a church door. . . .' The reverend gentleman then rather thoughtlessly associated the *Victoria*'s patrons with publicans and sinners, bringing down on his head a sharp rebuke from another correspondent! ' "Lord give us a good conceit of ourselves" is a favourite prayer of us Scotsmen—so at least some of our English critics tell us. Judging from the somewhat extraordinary letter of the Rev A. Hargreave in your issue of today it would clearly be a work of supererogation for him to offer up such a petition. With what cool assurance does this reverend gentleman class all those who differ from him as publicans and sinners! . . .' Doubtless the worthy but misguided divine was sufficiently deterred from his project by this and other letters, for there is no record of his having carried out his offer.

The *Victoria*'s passengers continued to be landed by the lifeboats for the rest of the season but there was nearly a serious accident on 22 August when one of them capsized in a choppy sea off Dunoon and several people were rescued from drowning. The other passengers insisted upon being landed at the pier and again scaled the barriers and broke the gates to gain access to the town. It was the last serious encounter of the year and the season ended with the whole matter of Sunday calls at Dunoon unresolved. The affair dragged on through subsequent years as Dawson Reid and his friends attempted to prove their case, but it was not until the end of 1901 that they finally won the point.

The *Victoria*'s Sunday sailings not unnaturally diverted attention from her ordinary weekday services. She did not sail on Wednesdays, which took the place of Sundays as the crew's rest day, but otherwise sailed daily at 9.30 from Glasgow to Skipness, calling at Kirn, Dunoon, Innellan, Rothesay and Tighnabruaich, arriving at 2.30 and returning from Skipness at 3.45. Dawson Reid gave great prominence to the cheap fares—from Glasgow to Skipness and back cost only 4s 6d in the saloon—and regularly advertised under the heading of 'Victorian Pleasure Cruising'. The Sunday cruise settled down as a trip round Bute during the summer of 1897. Other attractions included 'Piano and Organ on Board', and 'Special Rates' for societies and associations. A rare calling point at this date on the upper river was Erskine Ferry, where the steamer called on her Sunday sailings.

The year 1898 was more eventful as far as the railway fleets were concerned, both the South Western and the North British companies adding a new vessel. The former had considered building a duplicate of the *Jupiter* in September 1897 on the recommendation of Captain Alex. Williamson, and a tender was submitted by the Clydebank Engineering & Shipbuilding Co Ltd (successors to J. & G. Thomson) offering to build the vessel for £25,000. The offer was declined, however, after further discussion in October and no further reference to the acquisition of a new vessel appeared in the company's records until 14 June 1898 when the following minute was recorded :

> New Steamer : The chairman reported that the Clydebank Shipbuilding & Engineering Company (Ltd) had offered to sell to the Company a steamer 15 ft longer than the "Jupiter" which had been built for a Company who had been unable to raise the purchase price of the Steamer, and the question of the purchase was remitted to Mr Caird and the Chairman.

On 28 June another minute recorded :

> New Steamer : Mr Caird reported that . . . he had along with Captain Williamson inspected the Steamer offered . . . and the result of the inspection being satisfactory he had given instructions for the purchase of the Steamer at £27,000 on condition that certain alterations which he and Captain Williamson had suggested were carried out and that the Steamer be completed and ready for delivery not later than 4th proximo—the guaranteed speed to be 18½ knots on a four hours run. It was resolved that the Steamer be called the 'Juno'.

The origins of the *Juno* are to some extent mysterious, but the minutes quoted above support the long-accepted view that she was not designed for the Glasgow & South Western Railway. Of very heavy build, compared with the average Clyde steamer of her time, she had joggled plating, and her generally massive appearance suggested that she had originally been intended for cross channel work, possibly in the south of England, although confirmation of this is lacking. Her machinery was very powerful, the moving parts in particular being unusually ponderous. It was of the two-cylinder, compound diagonal pattern, steam being supplied by a large double-ended boiler. The new steamer replaced the *Neptune* as the Ayr excursion steamer and remained on these duties for substantially the whole of her career. Her design suited her well for service in the outer waters, while her great speed—she achieved a mean of 19.26 knots on trial on 5 July 1898—ensured her ability to sail to the furthest limits of the firth in the course of a normal day's working. Within three weeks of her entry into service, the *Juno* was in trouble; arriving at

Troon on 22 July, her engines failed to go astern, causing her to collide with the pier, damaging her stem. After repairs, however, she settled down to an otherwise respectable career.

Her older sister, the *Viceroy*, was involved in a much more serious accident on 14 May 1898. While crossing from Millport to Kilchattan Bay in broad daylight she collided with the Cardiff steamer *Godmunding*; both vessels sustained severe damage, the *Godmunding* being run ashore in a sinking condition. The South Western steamer's captain, Hugh McCallum, was held responsible and reduced to mate by the company.

The new North British steamer of 1898 was launched on 22 February, receiving the name *Kenilworth*. She was a repeat of the *Talisman*, but incorporated minor improvements and on trial on the measured mile on 30 May she attained a speed of 18.6 knots at an engine speed of 50 rpm. On 31 May she ran her official trials with a party of guests, sailing from Craigendoran to Dunoon and thence to Tighnabruaich, returning by the same route through the Kyles, dinner being served on the way home. Henry Grierson, chairman of the Steam Packet Company, made a speech in which he complimented A. & J. Inglis on the success of the new steamer and said that he hoped the time would come when they would be able to ask them to build a steamer with double engines, and of much larger capacity, for the North British Company—an intriguing hint that the company contemplated a change in its traditional policy of simple, cheap construction as early as the summer of 1898.

Although distinctly old-fashioned when she entered service, the *Kenilworth* was well finished. 'The after saloon is fitted with oak and handsomely upholstered with old-gold Utrecht velvet. The dining saloon, which is under deck forward, is fitted with solid mahogany, and the panels are of bronzed Lincrusta. Above there is a commodious smoke room. . . .' In appearance, of course, the new steamer was a copy of the *Talisman* and became her consort on the fast services to Rothesay. This released the *Redgauntlet* which, early in 1898, was refitted and a dining saloon installed in the lower deck aft. Thus equipped, she inaugurated a new daily cruise from Craigendoran round the Island of Bute on 1 June 1898. The *Redgauntlet* became popular on this service which, in time, became a regular Craigendoran excursion, but she was also used from 1898 on a variety of cruises.

WATER TUBE BOILERS

No new steamers were built for the Caledonian Steam Packet Company in 1898 but the *Meg Merrilies* was the subject of further reconstruction. Only two years before, the question of her reboilering had come up and a decision was minuted on 22 December 1896 that she should be given the second-hand Navy boilers from the *Marchioness of Lorne*, then herself about to be fitted with new boilers. The work on both vessels was carried out by A. & J. Inglis at a cost of £2,050 for the *Marchioness of Lorne* and £1,400 for the *Meg Merrilies*, during the early months of 1897. In the autumn, however, the board minutes recorded the receipt of a letter from the Haythorn Tubulous Boiler Syndicate, Ltd, the contents of which were not detailed, but evidently they contained an offer to supply the company's patent water tube boilers for use in a Caledonian steamer. The matter was dealt with more fully in a minute of 5 October :

> Amended offer from the Haythorn Tubulous Boiler Syndicate Limited was submitted, and Captain Williamson instructed . . . regarding additional conditions, and letter sent as follows :
> ### 'P.S. *Meg Merrilies*'
> With reference to your letter of 23rd ulto. I have now been instructed by my Directors to confirm the terms of your letter for the supply of two of your Company's Boilers for the P.S. 'Meg Merrilies' subject to the following additional terms :
>
> (1) In the event of the Board of Trade refusing to certify these Boilers or to renew their certificates from time to time during two years from the date of the Boilers being ready for work, your company will undertake to remove the boilers at their expense, and make good any damage to deck work which may be caused while removing the boilers.

Page 143: The tourist round: the *Lord of the Isles (II)*, as modified with extended promenade deck, disembarking passengers at Inveraray.

Page 144: Victorian heyday: Rothesay Pier in the late eighties—the first *Lord of the Isles* nearest the camera, and the *Adela*, of the Wemyss Bay fleet, in the western berth.

(2) In the event of my Company sustaining any loss or damage
or becoming liable to others in respect of loss or damage
caused by any accident arising from or in any way due
to these boilers . . . your Company will pay my Company
for any such loss or damage sustained by them, and will
also relieve my Company of any such liability incurred by
them. . . .

Acceptance of the conditions by the Haythorn Syndicate was
finally received in November 1897 but the new boilers were not
installed in the *Meg Merrilies* until April 1898. A minute of 15
November recorded that 'Correspondence with the Haythorn
Tubulous Boiler Syndicate, Ltd was submitted and discussed,
when it was agreed to pay £500, being the cost agreed upon for
these boilers, on the express condition that if the boilers are not
satisfactory at the end of two years from the date of these being
put on board the £500 is to be refunded'. The work of conversion
was done by William Denny & Bros at Dumbarton, and the
steamer's machinery was converted to compound expansion at
the same time. No difference in her appearance could be detected
save that there were now two waste steam pipes on her funnel.
The extra buoyancy resulting from the use of water tube boilers
was very marked, however, and within a short time the *Meg
Merrilies* was equipped with a fore saloon, which brought her
into line with all other members of the Caledonian fleet and
added immensely to her appearance.

The *Meg* was given a long trial on 21 December 1898, lasting
nearly six hours as she cruised around the firth. Steam was
supplied to the engines at a mean pressure of 125 lb per sq in
through a reducing valve, the boiler pressure being about sixty
pounds higher. The main results of the tests are recorded in
Captain James Williamson's book, his own summary being that
the boilers were highly efficient, but that the steam passing
to the engines was very wet. Further trials early in 1899 confirmed
this view. The results were evidently insufficient to justify the use
of water tube boilers in other steamers and unhappily the experiment which started amidst high hopes of success ended in recrim-

inations. In November 1899 the Caledonian Company demanded an extension of the guarantee period agreed on in 1898 so that the boilers could be repaired and tested, failing which they were to be removed by 15 April 1900. There ensued further correspondence between the Steam Packet Company and the boiler manufacturers, but the Caledonian remained adamant and made arrangements with A. & J. Inglis for the *Meg Merrilies* to be given new boilers of more conventional design at a cost of £1,230 failing an agreement with the Haythorn Syndicate. In February 1900 the steamer was disabled by an accident to one of the boilers and the Company had to charter the *Strathmore* for six weeks, debiting the expense to the Haythorn Syndicate. The *Meg* went to Blackwood & Gordon's yard at Port Glasgow and underwent repairs to the hull, and was thereafter reboilered, appearing in 1900 as a conventional vessel with Navy boilers.

Reference has already been made to the *Culzean Castle*, which had maintained a service from Fairlie to Campbeltown since 1895, but after the 1897 season she went up river, assumed her third name of *Carrick Castle*, and engaged in general excursion work. She was replaced on the Campbeltown station by Captain John Williamson's new *Strathmore*, which commenced a regular sailing on 30 May, calling intermediately at Millport, Lochranza, Pirnmill and Machrie Bay. This was almost certainly done by arrangement with the Glasgow & South Western Railway, whose board minutes referred to the matter of the Campbeltown service during June 1899 and a most interesting memorandum on the subject was prepared at that time (quoted in Appendixes). In the other private fleets in 1898 changes in services were of only a minor nature. Dawson Reid, however, returned to the limelight by floating a new company, The Glasgow Steamers, Ltd, and purchasing the old North British favourite, the *Jeanie Deans*, from her Irish owners. She replaced the *Victoria*, whose chequered career on her native firth had ended. Under a new name, *Duchess of York*, the *Jeanie* took up much the same pattern of services as her predecessor, including Sunday excursions, but the Skipness venture was not resumed. During this season the scenes at Dunoon were avoided, as the steamer was not advertised to call there.

A NORTH BRITISH COMPOUND

The North British Steam Packet Company had apparently discovered that Clyde excursion traffic was a lucrative trade following its experiment with the *Redgauntlet* during 1898, and on 22 September a minute recorded that 'on the proposal of the Chairman the manager was instructed to draw a specification of a new and improved steamer that would be suitable both for the ordinary Clyde traffic in connection with the North British Railway trains, and for excursion traffic'. Having prepared specifications, Robert Darling sent copies to the leading Clyde shipbuilders. 'After consideration it was resolved (on 20 October 1898) to accept Messrs A. & J. Inglis' offer of £24,200 for a steamer 235 feet by 26 feet by eight feet six inches moulded depth, with compound diagonal Engines and "Haystack" Water-tube boiler to give 2,300 indicated horse power, and the Secretary was authorised to conclude with Messrs Inglis accordingly.'

The new steamer was given the name *Waverley* at her launch at Pointhouse on 29 May 1899. She was undoubtedly one of the finest Clyde paddle steamers of the pre-1914 period and represented a considerable advance on any other vessel owned by the North British Steam Packet Company. She was twenty feet longer and three feet broader than her immediate predecessors, the *Kenilworth* and *Talisman*, and her saloons were considerably larger. The *Waverley* was designed with cruising in view and she was fitted with twin crank compound machinery of conventional type, which avoided much of the surging sensation associated with the larger steamers equipped with the single engine. The increased size of the *Waverley*, however, would almost certainly have required the use of compound engines in any case. She was the first Clyde steamer to combine a haystack boiler with compound machinery. This had been done successfully five years earlier when Hutson & Corbett so equipped the *Westward Ho!*

for Peter and Alexander Campbell's Bristol Channel fleet. There were sound advantages in the continued employment of the haystack. The North British fleet operated in very shallow water at Craigendoran and it was essential to use steamers with light draught. Haystack boilers were not as heavy as more modern designs, and naturally commended themselves in these circumstances, while of course they were unrivalled for rapid steam raising. The *Waverley* was therefore built with a modern, high pressure haystack boiler, with a working pressure of 110 lb per sq in.

The builders turned out one of their most elegant steamers in the *Waverley*. The fore saloon was for the first time in a North British steamer extended to the full width of the hull, and it was also extended forward of the mast, although the bow was open as in all her consorts. She was beautifully proportioned, with a single funnel, standard Inglis paddleboxes, and a small deckhouse amidships carrying the navigating bridge in the time-honoured position abaft the funnel. The *Waverley* went on trial on Saturday, 8 July 1899, with a party of officials and guests, and achieved the unusually high speed of 19.73 knots over the measured mile, which placed her in the same category as the *Juno* and *Glen Sannox* as one of the three fastest steamers on the Clyde. She cruised round Bute after her speed trial while the company on board were enjoying dinner and listening to the usual speeches. Unhappily, the proceedings were under a cloud, for Robert Darling, manager and secretary of the company for many years, had died on 31 May, and never saw his new flagship in service. He was replaced by J. W. Greenfield as manager and L. H. Gilchrist as secretary. The year of Darling's death was the high-water mark of the North British Steam Packet Company; with the introduction of the *Waverley* the fleet consisted of nine vessels, operating mainly in the upper firth but ranging as far as Campbeltown, Inveraray and Ayr on excursions. The *Waverley* took up service on the daily cruise round Bute on her arrival at Craigendoran, releasing the *Redgauntlet* for longer-distance excursions.

THE WRECK OF THE REDGAUNTLET

The *Redgauntlet* was involved in a spectacular accident on 16 August 1899 during a cruise round the Island of Arran, the third in a season's programme which also included trips to Ayr, Lamlash, Corrie and Lochranza, and a 'round of the lochs'. Leaving Craigendoran on a morning of clouds and a strong south-westerly breeze, the *Redgauntlet* encountered worse weather down the firth. As she sailed across from Millport to Arran and down Kilbrannan Sound, the wind occasionally reached gale force. Conditions became even worse as the steamer approached the south end of Arran, near Sliddery, and it was not appreciated that she was being blown closer to the shore than was safe in the circumstances. At 2.15 pm passengers in the dining saloon in the lower deck aft were appalled to hear a crashing noise as the *Redgauntlet*'s stern struck the Iron Rock ledges, a sunken reef about a mile off shore. At full tide these lay eight feet below the surface and the *Redgauntlet* would probably have sailed over them without harm; unfortunately, the tide was on the ebb, and she struck and was badly holed. There was a near panic as passengers rushed on deck. Captain McPhail did not immediately grasp the seriousness of the situation, but when a steward ran to the bridge shouting 'For God's sake beach her, or we'll all be drowned!' he instantly ported the helm and made to run the steamer ashore. He was just in time. The engineer, watching the sea pouring in as the *Redgauntlet* settled rapidly by the stern, was heard to shout 'We're gone!' as the water rose to the boiler, but at that moment the steamer grounded and an explosion was averted.

The *Redgauntlet* was aground fairly close to the beach and was in quite a safe position, but the spectacle of huge waves breaking over the submerged stern was sufficiently unnerving to those who had simply come on a pleasure trip. Many local inhabitants saw the disaster and came down to the beach and helped to bring the passengers ashore; there were only ninety-seven of them but the process took two hours to complete. Some walked several miles to Whiting Bay where they were taken aboard the *Waverley* which, in response to urgent telegrams, had been despatched from Craigendoran and arrived at 3 am. She left Whiting Bay an hour later and returned at eight o'clock in the morning. The *Lucy Ashton* was taken off her regular runs and sent to Whiting Bay the next day, and in the meantime the other passengers, who had been taken into practically every house in the district for the night, were conveyed to Whiting Bay by relays of brakes. It was a tedious process and the *Lucy Ashton* did not set off again for her home port until three o'clock in the afternoon.

Happily it was found possible to refloat the *Redgauntlet*. On 31 August the North British minute stated: 'The Manager reported that . . . he had arranged with the Clyde Salvage Company to lift the " Redgauntlet" and place vessel on Messrs A. & J. Inglis Slip at Pointhouse, Glasgow for the sum of £1,500, and that the vessel was duly floated and placed on Slip on 25th inst.' The damage had not been as extensive as had at first been feared and after repair and refurbishing the *Redgauntlet* duly resumed her duties. The Board of Trade inquiry into the accident found Captain McPhail in default, but in recognition of his prompt action in running the steamer ashore, thus ensuring the safety of those on board, his certificate was not suspended, but he was required to pay £20 towards the cost of the inquiry. The company reduced him to mate as a result of the affair.

The minute of 3 May 1900 recorded that the *Redgauntlet* had resumed her station on 16 April, and the total cost of repairs was noted as £5,737 14s 9d, while all claims by passengers had been settled for £336 13s.

The *Lucy Ashton* was eventually to become celebrated as one of the longest lived Clyde steamers, but she was in no sense a remarkable vessel at this period. Nevertheless, to those who knew her in much later years, it comes as something of a surprise to learn that the North British Company seriously considered disposing of her in 1899. A minute of 23 November 'agreed that in the event of a buyer making a firm offer for the "Lucy Ashton" such should receive . . . consideration'. On 3 May 1900 it was reported that she would need considerable repairs, but in view of the fact that negotiations for her sale were in hand, these were deferred. Several enquiries were made, and a Mr Constant, of London, made an offer of £2,000 for the steamer in July, but this was declined. Thereafter, there was a change of policy, and the *Lucy Ashton* was no longer offered for sale.

The last year of the old century brought a lull in construction in all Clyde fleets but David MacBrayne's famous flagship, the *Columba*, emerged rejuvenated after very extensive reconstruction. As built in 1878 she had been fitted with four double-ended Navy boilers, but they were replaced in 1900 by two large haystack boilers, sixteen feet in diameter, each with six furnaces. Being considerably lighter than the original installation, the result was that the *Columba* floated several inches higher in the water than before. The paddle wheels were thoroughly overhauled and the original wooden floats replaced by slightly larger steel floats. When the steamer resumed service it was found that her speed had risen from eighteen to well over nineteen knots, a remarkable achievement in a vessel then twenty-two years old. Internal refitting included the provision of a new companion way down to the main saloon, while the steamer was completely redecorated and other improvements made. Thus modernised, the *Columba* entered the best-known phase of her long career.

In appearance she and her older sister, the *Iona*, were definitely old-fashioned by 1900 but both ships were so stately that it was unthinkable so to describe them. That the *Columba*, the foremost vessel of the Clyde fleet of 1878, should remain so after a quarter of a century was remarkable, but she was no ordinary steamer. Our review of the closing years of the nineteenth century may fittingly conclude with the sight of this most imposing of Victorian pleasure steamers, even then a legend, renovated and improved for the new century's traffic and ready to face strong competition from a new and revolutionary form of passenger steamer.

CHAPTER SIX

THE NEW CENTURY—I:
THE TURBINE ADVENTURE
(1901-1912)

THE nineteenth century passed. Queen Victoria survived it by less than a month and her death at the end of January 1901 symbolised the end of much that was familiar and accepted. There came a new century and a new monarch, the first king that any save the oldest in the land could remember. It was a time of change, of experiment, and of new development in many fields, not least in marine engineering, and by a remarkable chance the Clyde found itself the proving ground for an invention of the most profound importance, the marine steam turbine engine, fitted into the first passenger vessel of the twentieth century.

THE PARSONS TURBINE

It was apparent to most steamer operators during the closing years of Queen Victoria's reign that the design of coastal paddle steamers had reached a peak. On the Clyde, vessels such as the *Juno* and *Jupiter* of the South Western fleet, and the North British *Waverley*, marked the limits within which ships could be built and operated without unreasonable expense, and the maximum speed attainable with any certainty lay in the $18\frac{1}{2}$-$19\frac{1}{2}$ knots range. The *Glen Sannox* was in a different category and in her case the limits of economy had really been exceeded. Built regardless of expense to recover the Arran traffic lost to the Caledonian in 1890, she had been more than successful in doing so but the cost of driving her at full speed was excessive even in the nineties and it is doubtful if she was ever fully extended in normal service. Her rival, the *Duchess of Hamilton*, although considerably slower, was nevertheless probably the better all-round steamer, having regard to availability and cost of running. Gradually it came to be appreciated that the class of ships of which she was representative, in the $17\frac{1}{2}$-$18\frac{1}{2}$ knots range, was best suited to Clyde conditions in general but while this applied to most routes there yet remained sailings to ports such as Campbeltown and Inveraray on which higher speeds were desirable so that the necessarily long running time might be curtailed. However, paddle steamer design offered no substantial improvement in this direction and matters stagnated for some years until in the late nineties a completely new form of marine engine was brought to a stage at which it appeared to offer a solution to the problem.

The steam turbine engine in its best known form was the invention of the Hon Charles Parsons, son of the Earl of Rosse. An engineer of considerable ability, he conducted his first experiments during the eighties, mainly in the field of electric generating plant and indeed one of the installations designed by him was used on board the *Duchess of Hamilton* in 1890 to provide power for her electric lighting system. The turbine offers many advantages over a conventional reciprocating engine, both mechanically and thermally. The principle on which it operates is relatively simple. Steam at high pressure is passed through consecutive drums of successively larger diameter, each containing alternate rows of fixed and movable steel blades. The latter are attached to a rotary driving shaft passing longitudinally through the drums, and the former to the turbine casing. As steam passes through the successive casings the movable blades are deflected, imparting a rotary motion to the shaft, and in so doing the steam expands and falls to a much lower pressure. It will be recognised that by converting heat energy, in the form of steam, into work through rotary motion, the turbine offers many advantages, amongst the more important of which is economy of size in relation to power, since the cylinders, connecting rods and ancillary machinery of a reciprocating engine are all avoided. Thermally, the turbine offers advantages, in that the range of expansion of the steam is much wider than in a reciprocating engine of comparable power and is thereby used more efficiently, with corresponding economies in fuel.

THE VIPER AND COBRA

It occurred to Parsons that his engine might be applied to the propulsion of ships and in January 1894 a syndicate was formed to build a vessel for experimental purposes. This was a small steam yacht, the *Turbinia*, which amply justified Parsons' view that the turbine could achieve very much higher speeds than existing forms of machinery. The point was dramatically emphasised on the occasion of the Jubilee fleet review at Spithead in 1897 when the *Turbinia* raced through the lines of warships at high speed, and no vessel present was able to overtake her. This demonstration of the potentialities of the turbine impressed the Admiralty sufficiently for an order to be placed in January 1898 with the Parsons Marine Steam Turbine Co, Ltd, of Wallsend-on-Tyne, which had been formed to exploit the invention commercially. The company had suggested to the Admiralty that a ship be built for the Royal Navy, and the order was placed for a torpedo boat destroyer with a guaranteed minimum speed of 31 knots. The hull and Yarrow-type boilers were sub-contracted to Hawthorn, Leslie & Co, the turbine machinery being constructed by the Parsons company. The ship was named *Viper*. She was a rakish vessel, with three large funnels and the turtle deck then usual in ships of her type. She was 210 ft long and 21 ft broad, and displaced 370 tons. The turbine engines were arranged to drive four propeller shafts, each fitted with two screws 40 in in diameter. The two outer, or wing, shafts were driven by high pressure turbines, from which steam passed to low pressure turbines driving the inner shafts. The inner shafts also carried astern turbines, and in manœuvring the destroyer live steam was passed directly into the low pressure ahead turbines or the astern turbines, the high pressure turbines on the wing shafts being temporarily isolated by non-return valves. For rapid ahead and astern manœuvring this arrangement was found to be very convenient, and the outer propellers idled in such circumstances. The *Viper* went on trials late in 1899 and surpassed all expectations by achieving a speed of over 37 knots.

A second turbine-driven destroyer was built during the same period but unlike the *Viper* she was a speculative venture by Armstrong, Whitworth & Co. This vessel, yard no 674, was a 'stock' ship, and the builders negotiated with the Parsons Marine Steam Turbine Company in 1898 with a view to fitting turbine engines. Like many of her contemporaries the ship was very lightly constructed and the boilers, as in the *Viper*, were of Yarrow pattern. The machinery was similarly arranged to that of the *Viper* but there were three propellers on each shaft. She was duly offered to the Admiralty who, significantly in the light of after events, insisted upon certain strengthening of the hull before agreeing to purchase at a cost of £63,500, and the name *Cobra* was given to the ship. Her external appearance differed considerably from the Hawthorn, Leslie destroyer, the most obvious divergence being four funnels instead of three. She was launched in June 1899 and ran her first trials in the following month, but an accidental collision caused damage which took some months to repair and it was not until June 1900 that further trials were undertaken. She was designed for a minimum guaranteed speed of 34 knots and achieved nearly a full knot more than this, but there was much dissatisfaction in Admiralty circles at her inability to reach her full potential, due to the fact that she could not accommodate enough stokers to maintain steam pressure at the highest speeds.

Some interesting information on the private negotiations which resulted in the first commercial application of turbine machinery to a vessel for Clyde service is obtained from the recorded discussion of a technical paper read by the Hon C. A. Parsons in 1901. Archibald Denny, a partner in William Denny & Brothers, Dumbarton, recalled that some years earlier he had heard a previous paper by Parsons and had been so impressed with the potentialities of the new engine that he had approached him to suggest its incorporation in a merchant ship. The Clyde railway steamboat owners appear to have been invited to sponsor a vessel of this kind although no formal record survives of such a proposal. Doubtless Denny's approach was informal and exploratory. However it was done, the response was discouraging and no support was obtained from the railways. In the meantime, however, Captain John Williamson, maintaining the long Fairlie–Campbeltown service with the dependable but pedestrian *Strathmore*, seems to have been considering the value of a turbine steamer quite independently and in due course he agreed to take the risk of inaugurating the new venture on the firth.

THE TURBINE SYNDICATE

Late in 1900 a syndicate was formed, the members being the Parsons Marine Steam Turbine Company, Ltd, which was to supply the machinery of the new steamer, Messrs William Denny & Brothers, the builders of hull and boiler, and John Williamson, who agreed to operate the ship personally during the 1901 season. It was stipulated, *inter alia*, that each member of the syndicate should contribute one third of the cost price of the new vessel (£33,000) and provide a further £267 towards initial working and running expenses. At the end of the season the steamer was to be sold as advantageously as possible or taken over by John Williamson, whose services during the first season were to be without remuneration.

An intriguing insight into the arrangements for placing the new steamer on the Fairlie and Campbeltown service is obtained from Glasgow & South Western Railway official records, in which the following minute, dated 22 January 1901, appears:

> Fairlie and Campbeltown Steamboat Service—Captain John Williamson, having represented that he had arranged with others for the building of a Steamer with Parsons turbine Engines and Propellers—It was agreed to guarantee his overdraft with the National Bank of Scotland Limited conditionally on Captain Williamson placing and maintaining the Steamer on the Fairlie route next summer.

It is evident from the above extract that in effect the South Western was backing an extension of its services into the parts of the Clyde prohibited to its own vessels under its parliamentary Act, and the new service on its inauguration gave connections not only at Fairlie but also at Prince's Pier, which became the turbine headquarters.

The experimental nature of the turbine venture demanded that no undue risks be taken with a completely new hull design and William Denny & Brothers fell back on the successful *Duchess of Hamilton* for a model, so that the new ship inherited her main dimensions, 250 ft in length and 30 ft in breadth. With a view to conversion to paddle propulsion in the event of failure, the promenade deck planking was noticeably designed for this contingency.

The new turbine vessel was given yard no 651. In concept she differed radically from the destroyers which preceded her for in their case the need was for light construction, high pressure boilers and outstanding speed. In the Clyde steamer however the call was rather for economy in fuel, involving a wider range of expansion than in the naval vessels. A lower pressure—150 lb per sq in—was proposed, so that a conventional double-ended boiler was used instead of Yarrow water tube boilers. Nothing approaching the speed of the *Viper* and *Cobra* was contemplated and a relatively modest twenty knots proposed for Clyde service sufficed to outpace any paddle steamer on the firth. A completely new layout of the turbine machinery involved the use of three shafts instead of the four employed in the destroyers. The central shaft was driven by the high pressure turbine, which exhausted into low pressure turbines, one on each outer shaft. It was stated that the high pressure steam expanded five times within the H.P. turbine and again twenty-five times within the L.P. drums before being finally exhausted to the condenser, so that the whole range of expansion was 125 times in comparison with the eight to sixteen times obtainable in a contemporary triple expansion reciprocating engine of the best design. As in the destroyers, astern turbines operated on the L.P. shafts, the central shaft idling while the ship was being manœuvred. The high pressure shaft carried one screw and revolved at 700 rpm at full speed, while the outer shafts, each fitted with two smaller screws, revolved at 1,000 rpm.

THE KING EDWARD AND QUEEN ALEXANDRA

The new steamer was launched by Mrs Parsons on 16 May 1901 and by royal permission the name given to her was *King Edward*. She was described as a handsome vessel and while she had not the grace of paddle steamers such as the *Duchess of Rothesay* and *Grenadier*, the clean, austere sweep of her hull, unencumbered by paddleboxes and sponsons, was undoubtedly attractive. In view of her work in the outer firth she was plated up at the bow, a feature that emphasised the neatness of her lines. The superstructure consisted simply of a small deckhouse supporting the navigating bridge, forward of a short flying deck supporting two smartly raked funnels and the two lifeboats. In view of the connection with Captain John Williamson the *King Edward* bore his colours—black hull, white saloons, and white funnels with black tops—and flew the well-known 'star and crescent' pennant of the old 'Turkish Fleet'.

The first steam trial took place on 14 June 1901 under the personal supervision of the Hon C. A. Parsons and John Williamson, and three days later the *King Edward* was on manœuvring trials, in the course of which she reached a speed of 18.66 knots as the mean of two runs over the measured mile at Skelmorlie in calm weather. She was then taken to Scott's yard at Greenock for hull cleaning and thereafter, on 24 June, ran a further series of trials, consisting of seven return runs over the mile, the best mean speed achieved on this occasion being 19.7 knots—a better result, but still below the expected 20 knots. On the following day she was slipped in the Pointhouse yard of A. & J. Inglis and the propellers were exchanged for a new set. The central propeller, 4 ft in diamater, was replaced by another of 9 in larger diameter, and the outer propellers, 2 ft 10 in in diameter, gave place to larger ones 3 ft 4 in in diameter. A third series of trials was then run on 26 June when, on a smooth sea, with light breeze, the *King Edward* achieved a mean of 20.48 knots over the measured mile, the fastest single run being done at 20.57 knots. The official trial took place in beautiful weather on 28 June 1901 when a party of invited guests joined the *King Edward* as she lay off Craigendoran, and thereafter sailed to Campbeltown. calling at Dunoon, Rothesay, Largs, Fairlie, and Lochranza. At the last-named, she was encountered by the *Duchess of Hamilton*, carrying a special party of members of the Institute of Naval Architects, and both steamers commenced a race down Kilbrannan Sound. In spite of the best efforts of the Caledonian steamer, the *King Edward* had no trouble in passing her, to the accompaniment of cheers from those on board. Peter Denny, of the builders, thereafter presided at dinner, in the after saloon and he and other speakers commented on the speed and smoothness of the ship. Such minor vibration as was noticeable was due to the action of the propellers in the water, according to Parsons— the technical problem of cavitation, to which he directed much attention during his subsequent career. Nevertheless, the steamer was markedly free of the rhythmic surging motion and vibration detectable in even the best of the paddle steamers, a feature that contributed much to her popularity with the public when she entered service. Despite forebodings in some quarters, the *King Edward* was found to be perfectly manageable at piers, although perhaps the flexibility of a paddle steamer was not evident, and in later turbine steamers attention was directed towards improving reversing power.

The machinery of the *King Edward* was her only experimental feature; in her general fittings she followed the best contemporary river steamer practice. She entered regular service on Monday, 1 July 1901 and became an instant favourite with the Glasgow public. The novelty of her design, the splendid weather of that summer, and additional traffic from which the Clyde in general benefited as a result of the Glasgow Exhibition of 1901 all combined to ensure the financial success of the turbine venture. The *King Edward* was advertised to leave Greenock (Prince's

Pier) daily at 8.40 am, and sailed to Dunoon and Rothesay before taking up the Fairlie connection at 10.20 and proceeding to Lochranza and Campbeltown, where she was due to arrive at 12.20 pm. Returning at 3 pm from Campbeltown, the passengers were timed to arrive at St Enoch station, Glasgow, at 6.18 pm. In connection with the steamer at Campbeltown, coaches undertook a 'Daily Excursion to the Shores of the Atlantic', at Machrihanish, a popular extension of the day's run for a modest additional cost. Within the Clyde area proper, the Glasgow & South Western links with the turbine syndicate were strengthened by an Isle of Arran tour by which passengers could travel outwards via the *Jupiter* from Prince's Pier to Brodick through the Kyles of Bute, travel by coach to Lochranza, and return via the *King Edward*, or the route could be reversed. An early experiment, attempted for the first time in July 1901, was the employment of the *King Edward* for evening cruises from Greenock. Trains left Glasgow at 6.5 pm and after a cruise of some two hours—'with music on board'—passengers arrived back in town at 10.25 pm. This was a very popular trip and rapidly became an established feature of turbine operation. In one way and another, therefore, the ultimate commercial soundness of the new venture was rapidly built up and so promising was the experience of the first weeks that the season was extended into and throughout September, after which the *King Edward* was laid up for the winter.

The remarkable achievement of the Turbine Syndicate was overshadowed in the late summer of 1901 by the unexpected loss of both pioneer turbine destroyers. On 3 August, while the *Viper* was engaged in naval manœuvres with the Channel Fleet, her captain failed to take sufficient precautions in soundings and navigation while in the dangerous waters round the Channel Islands. In thick fog, due to ignorance of her precise position the *Viper* was put aground on the Renonquet Reef. Fortunately the weather was calm and no lives were lost, but the *Viper's* light hull was torn open and she became a total loss in spite of salvage attempts.

Page 161 : The tandem compound diagonal engine of the *Caledonia* (1889).

Page 162: Engine rooms: (above) the single diagonal engine of the *Madge Wildfire*; (below) early type of compound diagonal, with four entablatures—PS *Neptune*.

Just over six weeks later, on 17 September, the *Cobra* was also lost, but in the most tragic circumstances. After the strengthening of her hull as required by the Admiralty she left the Tyne for Portsmouth Dockyard in command of Lt A. W. B. Smith. On board were her crew and many of the personnel of the Parsons Marine Steam Turbine Company, including Robert Barnard, works manager, in place of Parsons himself, who had been unavoidably prevented from accompanying his staff; a total of seventy-nine people in all. Early in the voyage she encountered heavy weather and off Flamborough Head was seen to break in two and sink rapidly, only twelve of her complement surviving. Most of Parsons' men were lost and for some months after the disaster the company's correspondence was heavily edged in black. It seemed for a time that turbine propulsion might be compromised by the almost simultaneous loss of the two destroyers but it rapidly became apparent that the cause lay in the inherent weakness of the ships' hulls, a defect not peculiar to the *Viper* and the *Cobra*, but general in ships of their type. During the court martial which investigated the *Cobra* disaster, the destroyers HMS *Crane* and HMS *Vulture*, on 10 and 14 October respectively, suffered severe buckling of their hulls in heavy seas and had to return to harbour. The fact was that attempts to increase the speed of these ships had been made at the expense of weight and, in the case of the *Cobra*, to quote the finding of the court martial, her loss was 'due to the structural weakness of the ship, that she was weaker than other destroyers, and that it is to be regretted that she was purchased into His Majesty's service'. She was described as 'the climax of a form of construction in which safety had been perilously sacrificed to lightness and speed'. Thus prematurely ended in tragedy the careers of these pioneer turbine ships, leaving the *King Edward* the sole survivor. The strictures on the naval ships applied in no way to her, and she continued her immensely successful opening season. The financial results were so good that the overdraft guaranteed by the Glasgow & South Western Railway was cleared out of revenue within the year and, in terms of the agreement between the members of the syndicate, the *King Edward* was duly acquired by a new company, Turbine Steamers, Ltd. All in all, the *King Edward* was an immediate and outstanding success, and Captain John Williamson proceeded at once to order an improved ship.

In a letter dated 1 October 1901, Wm Denny & Brothers offered 'to build for you an enlarged "King Edward" for the sum of £38,500 payable in the usual instalments, and at least one half in cash, the balance being in bills. As security for the latter the vessel to be mortgaged to Messrs. The Parsons Coy. and to us. . . . The machinery will be as in the "King Edward", but for about 15% more power. . . . It is expected the vessel will attain on the measured mile a mean speed of 21 knots, but as we have explained to you we cannot guarantee this. . . . The vessel to be delivered complete and ready for service not later than the 1st June 1902.' This formal offer was accepted in writing by John Williamson on 3 October 1901. In the meantime the Denny people had written to the Parsons Company as follows: 'We confirm our acceptance of your Mr Parsons verbal offer to supply his share of the machinery for an enlarged "King Edward" for the sum of £10,500. . . .' Following written confirmation from Wallsend, Denny then laid down the new vessel, yard no 670. Again John Williamson resorted to the Glasgow & South Western Railway for financial support and the following minute appeared in that company's records on 21 January 1902:

> New Turbine Steamer; The General Manager reported that the Bank overdraft of £2,500 referred to in Board Minute of 22nd January 1901 had been paid off. . . . He further reported that the Syndicate proposed to sell the Turbine Steamer "King Edward" and had ordered a larger Turbine Steamer to be called "Queen Alexandra". It was resolved to guarantee an overdraft to Captain John Williamson on account of the new Steamer to the extent of £2,500 with the National Bank of Scotland, Glasgow. Captain Williamson to give an undertaking that he will place and maintain the Steamer on the Fairlie & Campbeltown Route during ensuing Summer.

The choice of *Queen Alexandra* as a name for the consort of the *King Edward* was not only logical but popular. The second turbine steamer was rather larger than her predecessor, but, as the foregoing correspondence suggests, was otherwise an improved version of the pioneer ship. Like her, she had three propeller shafts, the central and outer ones being driven by high and low pressure turbines respectively, while the astern turbines drove the two wing shafts, each of which, as in the *King Edward*, carried two propellers. The double-ended boiler, constructed by the builders' associate, Denny & Co was, however, larger than the *King Edward*'s. Boiler pressure, however, remained at 150 lb per sq in in the *Queen Alexandra*.

The new turbine steamer was launched from the Leven Shipyard on Tuesday, 8 April 1902 by Miss Dorothy Leyland, daughter of Captain Christopher Leyland, who was a close associate of the Hon C. A. Parsons, a fellow-director of the Parsons Marine Steam Turbine Company, Ltd, and an important figure in the early history of the development of the turbine engine. After the launch James Denny gave some information about the *King Edward* and the performance of her turbine machinery. He said that if she had been fitted with the most modern type of twin triple-expansion engines instead of turbines, the best speed that could have been obtained from her would have been 19.7 knots compared with the 20.5 of the *King Edward* on trials, the difference representing a gain in indicated horse power of 20 per cent in the turbine's favour. But to drive the ship at all with conventional reciprocating engines would have been nearly impossible due to the additional weight and increased displacement and the attempt could only have been made at a much enhanced first cost and ruinous expenditure in fuel. Turning to the efficiency of turbine machinery, Denny then said that it had been found that this fell when the engines were worked below full capacity and that the efficiency of the *King Edward* had been found to increase in proportion to her speed. In the lower speed range, when the *King Edward* was working between 17 and 18 knots, corresponding to a power output of only 50 per cent, only then did ordinary reciprocating-engined steamers burn less coal per knot of speed.

It will be remembered that the first set of propellers used in the *King Edward* were discarded during her early trials in favour of larger screws, and the *Queen Alexandra* was fitted at first with the *King Edward's* original set. She ran her trials on the Skelmorlie mile on 19 May 1902 making six runs in a moderate sea, with a 20 knot wind. The best mean speed of two runs was 18.56 knots. She was thereafter docked at Scott's yard and cleaned before the second series of trials on 22 May, when she made twelve runs over the mile. On this occasion the weather was better, with a smooth sea and a light breeze. The results were very good, the *Queen Alexandra* achieving a best mean speed of 21.63 knots, although the best single run was at 21.82 knots, making the vessel the fastest on the Clyde.

The new ship sailed to Campbeltown with an official party of guests late in May 1902, outwards via the Kyles of Bute and Kilbrannan Sound and back by the east side of Arran, covering the distance between Campbeltown and Greenock in three hours.

In appearance the *Queen Alexandra* was very similar to the *King Edward* but the two steamers could always be easily identified, the more so while the *King Edward* was in original condition. In this form, her lifeboats were placed alongside the funnels but in the case of the *Queen Alexandra* there was a continuous boat deck extending from the navigating bridge, at the forward end, to the deckhouse over the saloon companion-way aft. This gave room for the boats to be placed further aft. The top deck thus made available to passengers was much appreciated and the *King Edward* was altered in time to give similar accommodation.

The *Queen Alexandra's* opening cruise was on 31 May, from Prince's Pier and Gourock, thence between the Cumbraes and up Loch Fyne. She took over the Greenock, Fairlie and Campbeltown service on the following Monday, 2 June, and the *King Edward*, thus relieved, was placed on a new route from Prince's Pier and Fairlie to Tarbert and Ardrishaig, subsequently extended to Inveraray, in direct competition with MacBrayne's *Columba* and *Iona*, and the Inveraray Company's *Lord of the Isles*. By so doing, the turbine fleet gave a Glasgow & South Western service into yet another quarter officially denied it by Parliament, although in this instance the connection between the steamboat owners and the railway company was admittedly less intimate than that linking the Caledonian Railway and its Steam Packet Company.

Although the *Queen Alexandra*'s trial speed was satisfactory, her propeller arrangement appears to have left something to be desired, and a series of trials was conducted into 1904, involving different propellers. Without any alteration, she was on trial on 3 October 1902, after which the *King Edward* set of screws was replaced by five new propellers, 4 ft 6 in and 3 ft $2\frac{1}{2}$ in in diameter respectively on the central and outer shafts, and further trials took place on 11 October, but with no improvement on the trial speed in May. In April 1903 there was an important change, involving the replacement of the four outer propellers by two new ones 4 ft 2 in in diameter, one on each shaft, there being no change on the central shaft. Another series of trials was run on 23 April, involving eight runs over the Skelmorlie mile in smooth sea conditions, with a light east wind, but the best mean speed achieved was only 20.75 knots. She was docked and her outer screws cut down to a diameter of 4 ft before the next trials, which took place on 8 May. In a choppy sea, with an eight-knot easterly breeze, the best mean speed was 20.7 knots, although the fastest single run had been done at 21.18 knots.

The final series of trials took place in the spring of 1904. On April 20, the *Queen Alexandra*, now fitted with Smyth's six-bladed propellers on the outer shafts, but retaining the original propeller on the centre shaft, went on the mile in a smooth sea, with light northerly breeze, and returned the still lower mean speed of 20.37 knots. In the circumstances the owners might have been expected to revert to the original set of screws, but instead a sixth change was made to 3 ft diameter outside screws of conventional design, the central one remaining unchanged. Thus equipped, the steamer completed her trials on 5 May 1904, again in calm weather, and after hull cleaning, and showed a distinct improvement in performance. The best mean speed rose to 21.12 knots, including a single run over the Skelmorlie mile at 21.24 knots. With these results the owners seemed content and the steamer suffered no further alterations.

A 'CALEY' TURBINE

The success of the *Queen Alexandra* was no less immediate than that of the *King Edward* and when other owners outside the Clyde realised that the turbine engine was commercially viable orders for new ships came rapidly to William Denny & Brothers and the Parsons Marine Steam Turbine Company. As might have been expected the initial orders were for cross channel steamers approximately similar in size and type to the successful Clyde steamers.

At this stage no further turbine vessels were projected for Clyde service but the Caledonian Steam Packet Company minute book contains the following reference dated 1 November, 1904 :

> A letter was read from Messrs. John Brown & Co., Ltd. to Sir James Thompson and plans submitted for a turbine steamer, which were carefully considered, but the offer declined.

As John Brown & Co, Ltd and their predecessors, J. & G. Thomson, had already built two excellent paddle steamers for the Steam Packet Company within the previous ten years, one wonders exactly what the proposed turbine vessel would have looked like, in view of what in due course was turned out for the Glasgow & South Western Railway by this yard. However, it can only be a matter of speculation, for Captain James Williamson and his directors were not ready to consider a turbine vessel in the fleet. Nevertheless, the question of the replacement of the *Galatea*, the original flagship of 1889, was then before the board. She had never been quite the success that the company had hoped for and by 1904 the necessity for reboilering the steamer brought to the fore her future role in the fleet. At a board meeting of 30 August 1904 estimates for reboilering were considered but instead of orders being placed it was decided to lay the ship up, and in May 1905 negotiations were entered into with a view to disposal. These fell through, but the *Galatea* was eventually sold to Italian owners in the summer of 1906.

In the meantime her replacement had been discussed by the directors on 11 July 1905 and a minute recorded: 'Captain Williamson instructed to get plans and an approximate cost for next meeting.' In view of the family connection between James and John Williamson it seems unlikely that the former was unaware of the performance details and other information relating to the *King Edward* and *Queen Alexandra*, and it is therefore hardly surprising that he should have approached William Denny & Brothers in connection with the order for a new Caledonian turbine steamer as well as John Brown & Co whose earlier proposals had been declined. On 14 August the Dumbarton company wrote to James Williamson 'offering to construct for you a Turbine Steamer of dimensions 250′ x 30′ x 10′ 6″ . . . for the sum of £30,000'. On the following day the Caledonian board met and a minute recorded:

The following were submitted, viz. . . .
Estimates for proposed new turbine steamer for the Ardrossan & Arran route—Committee recommend that a new steamer (Turbine) should be got, and that Messrs. Denny's offer for a 20 knot steamer be accepted. . . .

From this it will be noted that the dimensions of the *King Edward* were chosen for the new steamer in preference to those of the larger *Queen Alexandra*. In acknowledging the Caledonian acceptance of its offer on 17 August, the Denny company nevertheless expressed reservations on the proposed speed of the ship, stating that 'we regret to note that you have departed from the proposed premium for speed. For the sake of further data we should have liked to try for a higher speed than specified, while you would have had the chance of getting this at a much cheaper rate than if it had been aimed at in the first instance. We can hardly hope you will reconsider your decision on this point, but in case you should feel inclined to do so, we venture to point out that the machinery will be proceeded with shortly on dimensions corresponding to the 20 knots speed, and after that the question of speed may be regarded as finally settled.' When these points were discussed at a later board meeting the directors agreed to reconsider their decision and on 4 September Captain Williamson was able to write to Wm Denny & Bros stating that 'with further reference to your favour of the 17th ulto which I submitted to my committee at their meeting today, I was authorised to inform you that they will agree to your having a premium of £400 in consideration of your giving them half a knot over the guaranteed speed, and £800 if the speed is one knot over the guarantee, but any speed under these limits is not to be considered. I hope, therefore, you will earn the larger figure. . . .'

The new Caledonian turbine, Denny yard no 770, was originally intended to carry the name *Marchioness of Graham* in honour of Lady Mary Hamilton, the daughter of the twelfth Duke of Hamilton, and heiress of the Arran estates, whose wedding to the Marquis of Graham was due to take place in the early summer of 1906. A board minute of 6 February stated:

> It was suggested that the new steamer be named the 'Marchioness of Graham' but final decision was delayed until next meeting.

This took place a fortnight later when the decision was confirmed, but on 6 March it was further recorded that a 'letter from Lady Mary Hamilton was submitted regarding the name of the new turbine steamer, which is to be considered at next meeting', and on 20 March 1906 another minute stated baldly that 'It was resolved to name the new steamer "Duchess of Argyll".'

Thus, as a second choice, this most popular of Clyde steamers received her name. No details of the contents of Lady Mary's letter to the Caledonian directors were recorded, but it is almost certain that she drew their attention to the fact that her marriage would take place after the steamer was due to enter service— in fact, over a month later—and to bestow the name in anticipation would have been injudicious.

Although modelled on the *King Edward*, the Caledonian

steamer was a more elegant ship in many ways. The austere appearance of the pioneer turbine was accentuated by her stark black and white livery, but the Caledonian colours of pale pink and dark blue, with yellow funnels, softened the outline. The bow, too, was unplated above the mainrail and another feature that tended to break up the uncompromisingly severe appearance of the *King Edward* was the provision of large observation windows in the fore saloon instead of round ports. Finally, the *Duchess of Argyll* shared with the *Queen Alexandra* the feature of a top deck, albeit a short one, and her boats were placed at the stern, so that her profile was more satisfactorily balanced than in the *King Edward* before that vessel was altered. The cumulative effect of these modifications was to produce one of the most elegant and attractive of all Clyde steamers, a vessel to which the description 'swan-like' was frequently, and deservedly, applied.

Her machinery was similar to that of the pioneer turbine steamers but the *Duchess of Argyll* from the outset had only three propellers, 3 ft 8 in in diameter. These were exchanged as early as July 1906 for smaller ones, 3 ft 4 in in diameter. The boiler, of Scotch, or double-ended, type, as in the *King Edward*, was nevertheless of greater capacity, having a grate area of 170 sq ft and heating surface of 5,322 sq ft. For ease of manoeuvring inside Ardrossan Harbour, the *Duchess of Argyll* was equipped with a bow rudder. There were also refinements in the engine room; those who recall the *King Edward* will remember the starting platform placed on the lower deck amidst a maze of steampipes leading to and from the boiler and turbines, and the *Queen Alexandra* is understood to have had a similar arrangement. But in the *Duchess of Argyll* the control platform was at main deck level, as in the paddle steamers, and the control wheels and gauges were far more neatly disposed in a layout which became virtually standard in later Clyde turbine steamers.

The *Duchess of Argyll's* first trials took place on the Skelmorlie mile on 4 May 1906 in a rough sea and stiff breeze, when she achieved a mean speed over two runs of 20.9 knots. Four days later in calmer conditions, she 'ran the lights'—Cloch to Cumbrae—and achieved a mean speed of 21.11 knots with a fastest single run of 21.65 knots. She was therefore fractionally slower than the *Queen Alexandra* on the basis of trial results but for all practical purposes there was nothing to choose between the two steamers. In any event the builders had substantially exceeded the guaranteed speed and received the maximum premium of £800 under the terms of the contract. The *Duchess of Argyll* took up service on the Ardrossan–Arran station on 19 May and for the first time showed the *Glen Sannox* a clean pair of heels on the Brodick run. The new turbine was commanded by Captain McKellar, while Captain Allan Macdougall, formerly of the *Duchess of Rothesay,* took over the *Duchess of Hamilton* in succession to the celebrated Robert Morrison, who retired at this time. The *Duchess of Hamilton* sailed thereafter on excursions and railway connections from Gourock.

THE ATALANTA

The summer of 1906 saw the introduction of a fourth Clyde turbine steamer, built for the Glasgow & South Western Railway. This vessel was quite outside the main stream of development on the Clyde itself. For her origins we must return to 1905 when a South Western Steam Vessels Committee minute of 19 September stated as follows :

> Proposed new steamer. Read letter from the Marine Superintendent to the General Manager dated 15th inst., recommending that a new steamer should be built. The Marine Superintendent was instructed to obtain alternative tenders for a Turbine and a duplicate of the steamer 'Mars'.[1]

The Committee met again on 17 October and a minute recorded :

> New Steamer : Submitted Drawings of a proposed Turbine Steamer and also a duplicate of the 'Mars' with certain improvements :—Authority was given to accept the offer of John Brown & Company, Limited for a Turbine Steamer as per their offer of 3d. October at £21,000.

[1] Built 1902—cf. Ch. VII.

The background to this order is of considerable interest. During 1905, John Brown & Co, Ltd had the Cunard liners *Caronia* and *Carmania* under construction at Clydebank and while the former was fitted with quadruple expansion reciprocating machinery it was decided to equip the *Carmania* with Parsons turbines. At the same period the decision was taken to build the still larger *Lusitania* and *Mauretania* as turbine ships, and John Brown & Co received the order for the first-named. In order to study the constructional and other problems involved, the shipbuilders assembled a small set of triple turbines on an experimental basis, following the standard layout of Denny/Parsons steamers for the Clyde and cross channel services. Naturally at a later stage these engines would have become surplus to requirements and it is *possible*—no more—that the offer to build a turbine steamer for the Caledonian Steam Packet Company in November 1904 envisaged the incorporation of the experimental machinery. Whatever the truth of the matter, the engines were eventually used in the new Glasgow & South Western steamer of 1906, to which the name *Atalanta* was given in February of that year.

This steamer has always been regarded as the least successful of all Clyde turbine-driven vessels but in fairness to the ship and her builders it must be realised that she was never intended to compete on terms of equality with the others. She was shorter by a good margin—only 227 ft—and broad in proportion to her length—30 ft—so that great speed could not be expected of her. The guaranteed trial speed was in fact only $17\frac{1}{2}$ knots and it is apparent from this and other evidence that she was intended to work turn and turn about with the other steamers of the South Western fleet as a general-purpose vessel—certainly not as a flyer in the *Glen Sannox* tradition. Unlike all the other turbines, steam was supplied by two Navy boilers with uptakes into a single funnel, so that even her appearance differed from that of the *King Edward* and her descendants.

The *Atalanta* was due to have been launched on Saturday, 21 April 1906 but the weather was unfavourable and a simple naming ceremony took place instead. She was launched without further ceremony on the following Monday, virtually completed, and took up her station early in May. South Western records contain no reference to her trial performance but a minute of 16 April 1907 implies that she had not been entirely satisfactory at first :

> 'Atlanta'. Read letter from the Marine Superintendent reporting that certain alterations had been made on this steamer at the Builders' expense and that on trial she had more than attained the guaranteed speed of 17½ knots.

Whatever the problems, it is evident from the minute that they had been successfully dealt with by the builders.

One of the *Atalanta*'s earliest duties was a special run from Ardrossan to Brodick on 27 June 1906, bringing the Marquis and Marchioness of Graham home from their honeymoon. *The Glasgow Herald*'s description is worth quoting in full :

> The Marquis and Marchioness, who left London on Tuesday night, arrived in Glasgow early this morning, and after a brief halt at St Enoch Hotel they journeyed with the 10.20 express to Ardrossan, travelling in a special saloon carriage. Mr Cockburn, superintendent of the line, travelled with the train to Ardrossan. Their arrival there, as their departure from the city, was witnessed by large crowds. The pier station at Ardrossan was lavishly decorated. At Ardrossan they joined the turbine steamer *Atalanta*, which had been placed at their disposal by the Glasgow & South Western Railway Company. Captain Williamson, manager of the Company's steamers, travelled on the *Atalanta* to Brodick. The steamer, which is under the command of Captain Gregor, was prettily beflagged, and a special boudoir, covered with a red awning and adorned with roses on each side, was erected on the deck for the use of the Marchioness. The crew wore their white jerseys, which added to the brightness and colour which pervaded the whole vessel.

The *Atalanta* became identified with no particular route in her first few seasons but there is a reference to her being on the Stranraer excursion sailing in 1907, although her suitability for the potentially stormy waters of the outer firth might be questioned.

Like most of the company's paddle steamers she was plated up to promenade deck level at the bow and with relatively small windows and ports forward she would have been perfectly sea-worthy, but almost alone amongst Clyde steamers of the period she had a reputation for liveliness in even moderate seas. A feature which made her unique in her own fleet was the navigating bridge, at last placed forward of the funnel, while aft there was a short top deck on the purser's office.

The pioneer turbine steamer *King Edward* appeared in new guise in 1906. Now owned, with the *Queen Alexandra*, by the new company, Turbine Steamers, Ltd, in which John Williamson had a substantial interest and was managing director, she had been taken in hand by her builders in the winter of 1905-6 and altered by the provision of a top deck extending aft over the saloon staircase but not forward to the bridge. The boats were moved to the after end of the new deck, allowing the funnels to be seen properly from a broadside position. A cloakroom was added, together with a smokeroom on the main deck. In her new form the *King Edward* was considerably improved not only in her appointments but also in appearance. Her first sailing as altered was advertised as a 'Grand Saturday afternoon opening cruise to Kilbrannan Sound' on 12 May 1906, after which she took up the Campbeltown run as usual until the arrival of the *Queen Alexandra* later in the month, when the *King Edward* reverted to the Inveraray station.

When increasing costs eventually brought a rapprochement between the Caledonian and Glasgow & South Western com-panies in the winter of 1907-8 culminating in a formal working agreement in 1908 and subsequent seasons with the object of eliminating wasteful competition, one of the first steamers to be affected by retrenchment and rationalisation of Clyde services was the *Duchess of Argyll*. She and the *Glen Sannox* provided between them an Arran service which by the standards of sixty years later was unbelievably lavish. Even for the halcyon years it was too good, and certainly too costly, and the Ardrossan–Arran routes were the first to be combined, the South Western and Caledonian companies thereafter working it alternately in summer, the steamer connecting with the trains of each company at Montgomerie and Winton Piers, Ardrossan. In 1908 the *Duchess of Argyll* suffered the indignity of being laid up and it is not inconceivable that had she not been the crack ship of the fleet she might have been sold. Eventually the *Marchioness of Bute* was disposed of to the Tay, and she was followed by other disposals in later seasons, but even so the *Duchess of Argyll* was not fully employed. She ran the combined Arran service in 1909 but in March 1910 came an unusual development when she was altered by having her bow plated up to promenade deck level and her large observation windows forward replaced by ports. In this form, which became permanent, she was used as a relief vessel on the Larne–Stranraer service in 1910 and 1911, in the latter year because of the breakdown of the *Princess Maud*. Apart from differences in detail, and company liveries, the *Duchess of Argyll* now resembled the *King Edward* very closely.

FIRE ON BOARD

The *Queen Alexandra* was a very popular steamer and the sailing public doubtless anticipated for her a long and honourable career such as was being enjoyed by David MacBrayne's *Iona* and *Columba*. Her withdrawal from service at an unexpectedly early stage was therefore a general surprise. In the early morning of Sunday, 10 September 1911, as she lay at her coaling berth in the Albert Harbour, Greenock, it was discovered that she was on fire, due to unknown causes, and by the time the fire brigade arrived the vessel was well alight, her sides being so hot that firemen could not approach closely enough until the hull plating was damped down. Two hours passed before the outbreak was under control and when the fire was finally put out the ship was seen to be severely damaged. The main and upper decks were burned through, while the promenade deck was badly damaged. On the main deck the bar, confectionery stall and saloon were all destroyed, and the dining saloon below was also gutted. The entire after part of the *Queen Alexandra* was in fact destroyed, and the amount of the damage was estimated at between £3,000 and £4,000.

The immediate problem was to continue regular turbine services. By a fortunate coincidence the *Duchess of Argyll* had been chartered for the following day to replace the *Queen Alexandra* on the Campbeltown run, while she took a special party on a cruise from Greenock. The *Duchess of Argyll's* charter was extended for a period and she was named in John Williamson's advertisements until 25 September, when the Inveraray sailings were withdrawn and the *King Edward* took over the Campbeltown sailings until the end of the month, when she too came off for the winter.

Although the fire at Greenock had not damaged the *Queen Alexandra* beyond repair John Williamson eventually decided that it would be better to replace her with a completely new steamer. Despite her success she had been notably deficient in reversing power, a fault common to most of the early turbine vessels. Williamson immediately began negotiations with William Denny & Bros for the construction of a new ship and on the undertaking of Colonel John Denny that it could be available for service by 1 June 1912 the directors of Turbine Steamers Limited decided to place the order and sell the *Queen Alexandra*. She was refitted and sold to the Canadian Pacific Railway for service at Vancouver, sailing thence via Cape Horn. Under her new name of *Princess Patricia* she sailed for many years until her final withdrawal in 1937.

On 7 October 1911 John Williamson wrote to William Denny & Bros:

> Referring to our various conversations and your offer to replace this steamer for the sum of thirty nine thousand pounds . . . I have been instructed by my Directors to accept your offer, and that the price of the new steamer is not to exceed this amount.
>
> Steamer to be delivered and ready for passengers on the 1st of June, 1912, as personally promised by Col. Denny, which promise was the means of my being able to carry this sale of 'Queen Alexandra' through, and the ordering of the new steamer, so I

hope you will do everything in your power to turn out as good a ship as the 'Queen Alexandra' proved to be.

As you know, the reversing power of the 'Queen Alexandra' was very defective, and the new steamer will require to have more reversing power than the 'Duchess of Argyll', and I rely on you making a much finer job in the Engine Department than the 'Duchess of Argyll'.

A NEW QUEEN ALEXANDRA

The new steamer, which was given the yard no 970, was in many respects a repeat of her predecessor, the main improvements, as suggested by the official correspondence, being in the engine room. She was named by Captain Leyland's ward, Miss A. M. Chetwynd, on 8 April 1912, but the actual launch had to be postponed due to bad weather. Once again the name *Queen Alexandra* was selected, no doubt in view of the goodwill built up by the original vessel.

The design of the turbine machinery in the new *Queen* followed the traditional pattern of three screws, a high pressure turbine driving the centre shaft, with low pressure turbines on the outer ones, the reverse turbines also driving the latter. The specification of the *Duchess of Argyll's* machinery was closely followed, as this was a considerable advance on the engines of the first *Queen Alexandra*. Reversing power was substantially improved; in the first *Queen* the astern turbines included six expansions, each of four rows of blades, whereas her successor's astern turbines had seven expansions, each with six rows. Reversing and manœuvring capacity was thereby much enhanced while the provision of a bow rudder, a feature not previously employed, added further to manœuvrability. It was estimated that the new *Queen Alexandra's* reversing power was about fifty per cent greater than that of the first ship, and on trial she attained an astern speed of $12\frac{1}{2}$ knots. Her full ahead speed, when on trials on 18 May 1912, was $21\frac{1}{2}$ knots, so that she was the equal of the first *Queen Alexandra*. It will be recalled that the high pressure shaft of the latter ship had a propeller larger in diameter than the

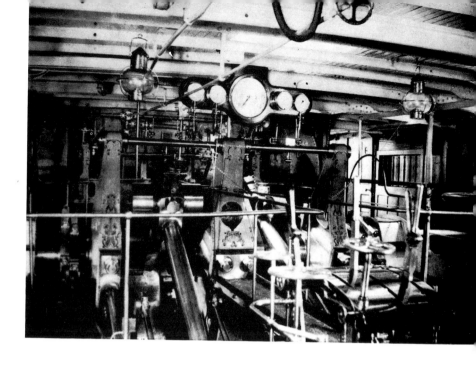

Page 179: Engine rooms: *(above)* J. & G. Thomson's compound engine for the *Jupiter*; *(below)* the John Brown engine of the *Duchess of Montrose*.

Page 180: Sweet Rothesay Bay: The *Viceroy* amongst the racing yachts, about 1906. The *Columba* and *Duchess of Montrose* (background left and centre) are also leaving the bay, while one of the Caledonian *Marchionesses* of 1890 (background right) approaches Rothesay from the Kyles of Bute.

outer ones, but in the new steamer all three screws were 3 ft 8 in in diameter and revolved at 800 rpm at full speed. Boiler pressure was slightly higher in the new steamer, at 155 lb per sq in.

Amongst various improvements the new *Queen Alexandra* was equipped with a telemotor for operating the steam steering gear, so dispensing with the usual rods and chains. This was the first example in a Clyde steamer. As in the *Duchess of Argyll* the engine room starting platform was at main deck level, with a neater arrangement of controls than in the earlier steamers. Externally the new ship differed only in detail from the older one, the principal distinguishing feature being that the navigating bridge of the later steamer was rather higher above the top deck than previously, and by this she could readily be identified.

The new turbine's first public sailing took place on the King's Birthday Holiday, 23 May 1912, when she sailed on an opening cruise from Greenock and Gourock to Campbeltown, going by the Kyles of Bute and returning via the Garrochhead. Under the command of Captain Angus Keith, who had also served in the old *Queen Alexandra*, the new vessel took up regular daily service on the Campbeltown route from Prince's Pier and Fairlie on 3 June, calling at Lochranza, Pirnmill and Machrie Bay. It may be remarked at this point that the celebrated excursion 'to the shores of the Atlantic' at Machrihanish, in connection with the turbine steamers, had been conducted since 1907 by the trains of the narrow gauge Campbeltown and Machrihanish Light Railway instead of horse-drawn coaches. This line, the only example of its kind in Scotland, had originally been a colliery railway but it was improved and adapted for passenger traffic in 1907. Captain John Williamson became one of its directors.

With the arrival of the second *Queen Alexandra* the first stage of turbine development on the Clyde was completed, and the first world war and its aftermath delayed further progress for nearly a decade and a half. The four steamers in service in 1914 were the *King Edward* and the new *Queen Alexandra*, of Turbine Steamers, Ltd, and the Caledonian Steam Packet Company's *Duchess of Argyll*, these three belonging to the original Denny/

Parsons type, while the remaining steamer, the Glasgow & South Western Company's *Atalanta*, was of rather different design, save for the layout of her machinery, and she was intended for more general duties.

The significance of the pioneer turbine steamers *King Edward* and *Queen Alexandra* was profound, and the advent of these vessels had a strong influence on naval architecture. Had it not been for their instant success, turbine development might well have been delayed for several years; as it was, progress was rapid after their introduction and the faith of the Dennys and John Williamson in the invention of Charles Parsons was amply justified. The subsequent triumph of turbine machinery over other types, and in particular its application to ocean liners, was directly attributable to the pioneering work on the Firth of Clyde in 1901, and the *King Edward* will always be regarded as one of the most notable ships of her type ever built.

TURBINE STEAMER

FOR

DUNOON, ROTHESAY, LARGS

FAIRLIE &

CAMPBELTOWN

CHAPTER SEVEN

THE NEW CENTURY—II
(1901-1914)

R UINOUS competition amongst the railway companies at last brought retrenchment and rationalisation during the first decade of the twentieth century. The Edwardian midsummer saw Clyde steamer services at their finest, but the public image of the railways was tarnished by the Millport affair of 1906. One of the Clyde's rare disasters caused the loss of the *Kintyre* in the following year and it is possible in retrospect to trace the subsequent numerical decline of the Clyde fleet as a whole from that date. Thereafter new construction fell short of disposals and the reduction of railway services before the Kaiser's War was counterbalanced only partially by a return of the popularity formerly enjoyed by the up river steamers.

THE LOCH ECK ROUTE

Change was the keynote as the new century dawned. Prospects of change and expansion were welcomed in many quarters and nowhere more warmly than in the City of Glasgow whose great International Exhibition dominated the year 1901. In retrospect it is possible to regard this as the high-water mark of Glasgow's importance as an industrial and commercial centre—the proud title of 'The Second City of the Empire' was well deserved. Proud of its industries, confident in itself, the city did not permit the old queen's death to inhibit public rejoicing as the Duke and Duchess of Fife arrived in Glasgow to declare the Exhibition open on 2 May. It was situated in Kelvingrove, on the site of its forerunner of 1888, but was on a much grander scale. The west of Scotland enjoyed one of the best summers in living memory and attendances at the Exhibition broke all records. Visitors flocked to the city from England and Europe and on the Clyde special yachting regattas drew thousands to the coast, where the steamers experienced the most successful season for years.

The advent of the turbine *King Edward* stole the limelight on the firth but her commercial success was matched by the experience of other owners. Both the *Columba* and *Lord of the Isles* appeared in service earlier than usual and did not retire to winter quarters until late in the year. The Loch Eck tour, operated in connection with the *Lord of the Isles* and the little gondola steamer *Fairy Queen* on Loch Eck itself, was improved by the provision of a spanking new four-in-hand char-à-banc which drove through the streets of Glasgow on its way to Dunoon on 6 May, in charge of a driver and guard in red coats and white satin hats. The vehicle carried thirty passengers on a sloping bank of seats, of which some were reserved for smokers. It was used to convey tourists from the *Lord of the Isles* from Dunoon to Loch Eck, returning with those who had chosen to disembark at Strachur, on Loch Fyne. The extensive nature of the services connecting with the *Lord of the Isles* is a matter for astonishment in the declining years of public transport. Leaving Glasgow at 7.20 am, the steamer called at Prince's Pier to embark passengers from as far afield as London (St Pancras) and the Midlands of England, by the Midland Railway's overnight service, as well as local passengers from Glasgow and South Western stations. Calling also at Gourock, she took on board Caledonian Railway passengers and English tourists who had travelled by West Coast route from Euston, while at Dunoon there was a fresh influx from the Craigendoran steamer conveying passengers from Edinburgh and East Coast towns, in addition to North British traffic.

All of the railway fleets enjoyed a busy year, but no new tonnage was added. Indeed, both the North British and the South Western suffered reductions, the former by the sale of the *Diana Vernon* to Captain Lee, of Brighton, in March, and the latter by the disposal of the old Lochlong & Lochlomond veteran, the

Chancellor, to a Spanish company, La Herculina Ferrolana, about the same time.

CALEDONIAN CHANGES

There was some activity in the Caledonian Steam Packet Company's fleet as the first generation of steamers came in for reboilering. Thus, the *Marchioness of Bute* was taken in hand by the Clyde Shipbuilding and Engineering Co in the spring and her boilers renewed for the sum of £2,800, while the *Marchioness of Breadalbane* went to A. & J. Inglis at Pointhouse in November for similar rebuilding, although in her case the cost was £3,250. The directors also agreed to have the *Ivanhoe*'s haystack boilers retubed in October, 1901 at a cost of £200, the work being done by Rankin and Blackmore. In August, the question of providing the *Galatea* with new boilers first arose, and Captain James Williamson was instructed to prepare a report on the project, which was also intended to include other alterations to the steamer. This appears to have brought into the open a feeling that the *Galatea* was no longer suited to the company's requirements, for a minute of 3 September 1901 recorded that 'the Secretary was instructed to try and obtain a firm offer of £10,000 for the "Galatea".' The former flagship had never really been the success that had been expected, her speed in particular being rather disappointing, and she was not as well designed as the *Duchess of Hamilton*. The Caledonian directors seem to have decided to spend no further large sums on her, and her story from the turn of the century until disposal is one of continual repairs to keep her sailing, and of efforts to sell the ship.

In October 1901 the directors turned their minds to ordering a new steamer, and tenders were invited from six of the Clyde yards. Although the *King Edward* was by this time a known success, the time was not yet ripe for the general introduction of turbines. The Caledonian board specified a paddle steamer 210 ft long by 24 ft broad, and the contract speed was as low as sixteen knots. On 29 October, the company's minutes recorded accept-

ance of John Brown & Co, Ltd's offer to build the ship for
£19,572, this company having taken over the shipyard and business of J. & G. Thomson, Ltd, at Clydebank. Plans were submitted to the board in December 1901 while on 4 February 1902
'it was agreed, subject to the approval of Her Grace, the Duchess
of Montrose, that the new steamer should be named "Duchess of
Montrose".' She was launched during early May and ran her
trials on 4 June 1902, achieving the contract speed.

The *Duchess of Montrose* was a smart little steamer, generally
resembling the *Duchess of Rothesay*, but the use of Navy boilers
instead of a 'double ender' in the older steamer caused the funnel
to be placed well forward. It was a very vertical funnel, and the
Duchess of Montrose was never the beauty she might have been.
It is strange how minor details can spoil the appearance of an
otherwise attractive ship, and just how close the *Montrose* had
come to perfection was emphasised in the following year when
she was joined by her consort and quasi-sister, the *Duchess of
Fife*. Despite the reservations about her good looks, the *Duchess
of Montrose* was a useful and economical vessel for general work
in the Caledonian fleet. The triple expansion type of paddle
engine used first in the *Marchioness of Lorne* was reintroduced;
it will be recalled that this form of engine incorporated four
cylinders—two high pressure, an intermediate, and a low pressure—arranged in tandem to drive two cranks. Captain James
Williamson himself recorded that this type of machinery was more
efficient than the conventional two-cylinder compound on long
runs, and that its greatest advantage lay in manœuvring at piers.
As regards the machinery of the *Duchess of Montrose*, John
Brown & Co, Ltd produced a beautifully made set of engines,
rather smaller than those of many of the large paddle vessels built
in the nineties. They drove very small paddle wheels at a fairly
rapid rate, the intention being to save wear and tear by avoiding
the use of unduly large wheels—precisely the opposite of Captain
William Barr's suggestion of huge wheels for the *Edinburgh
Castle*. Unfortunately, the diminutive paddleboxes detracted
further from the steamer's appearance, and the *Duchess of Mont-*

rose was always readily recognisable by this feature and the straightness of her funnel. In all other respects, however, she incorporated the best features of Caledonian practice—saloons fore and aft, open bow under the promenade deck, and well-designed accommodation on deck and below.

The arrival of the *Duchess of Montrose* finally meant the end of the road for the *Meg Merrilies*. The two steamers were used together in the Caledonian fleet in the summer of 1902 and it was not until 12 August that the records included reference to the sale of the *Meg* for £5,000. She was purchased by the Leopoldina Railway Company of Brazil for service at Rio de Janeiro, and our last glimpse of this, the first Caledonian steamer, is of her leaving Gourock during September 1902, specially fitted with a protective turtle deck for her Atlantic voyage, and heavily boarded up round saloons and paddleboxes. The *Meg Merrilies* was one of the unluckiest vessels ever to sail on the Clyde in the sense that she gave little satisfaction to any of her owners but despite all her misfortunes, there is good evidence that she was a popular steamer with the travelling public who at any rate were sorry to see her go.

The Glasgow & South Western Railway added another steamer to its fleet in 1902, as a replacement for the *Chancellor*. The company's minute of 20 August 1901 is of much interest, revealing as it does how closely the new vessel approximated to the *Duchess of Montrose*. 'An application by the Marine Superintendent for a new steamer . . . was submitted along with an offer by John Brown & Co. (Ltd) Clydebank to build a steamer according to specification with a speed of 16 knots, for the sum of £17,500. It was Resolved to recommend the Board to accept the offer.' Plans were submitted to the company in October and in December it was agreed to name the new steamer *Mars*. She was ten feet shorter than the *Duchess of Montrose*, and slightly broader, and was driven by a compound diagonal engine of normal type, steam being supplied by Navy boilers. Unlike her Caledonian rival, her bow was plated up, and the navigating bridge was abaft the funnel, so that the *Mars* was in general appearance a smaller edition of the *Jupiter*. She shared with the *Duchess of Montrose*—but with no other vessel in either fleet— the feature of very small paddle wheels and paddleboxes, the adoption of which was possibly attributable to the persuasion of the shipbuilders. The *Mars* was launched on 14 March 1902 and she entered service on the Prince's Pier—Rothesay—Kyles of Bute—Ormidale service in June.

No new steamer was added to the North British Steam Packet Company's fleet in 1902, but the company gained the equivalent of the *Mars* and *Duchess of Montrose* by the thorough reconstruction of the *Lucy Ashton*. This vessel had suffered a major machinery breakdown during the 1901 season and the North British Company decided to have her re-engined and reboilered. She was brought to Pointhouse, where A. & J. Inglis built and fitted a new twin crank compound diagonal engine and a high pressure haystack boiler. These alterations were done in 1902 and in the following year the deck saloons were extended to give additional covered accommodation. The result was to provide the North British with a completely modernised, economical little steamer for all-the-year-round service. It was virtually the final act of the Steam Packet Company to convert the *Lucy Ashton*, however, for in October 1902 it was dissolved and its assets, including the steamers, vested directly in the North British Railway. The effect, of course, was to impose upon the Craigendoran fleet the statutory restrictions which already inhibited the Glasgow & South Western Railway in its Clyde traffic, namely, in respect of sailing to certain ports on the western fringes of the Firth of Clyde, but in this step it is possible to detect a desire on the part of the North British to cut expenses by eliminating long distance cruises and concentrating the available fleet on more remunerative short runs. The new policy also allowed a vessel to be disposed of, and the choice fell on the *Lady Rowena* which, after only a dozen years, was sold to Italian owners in 1903.

David MacBrayne's winter steamer, the *Grenadier*, was the subject of a heavy overhaul in the winter of 1902-3, during which her original Scotch boilers were removed and replaced by haystacks. The change in her appearance was immeasurably for the better. The *Grenadier* had always been notable for the fine lines of her hull, but two thin funnels placed close to the paddleboxes marred her good looks. After the alterations she emerged as one of the acknowledged swans of the firth. New and larger funnels set further apart, and well raked, gave her a beautifully balanced outline, set off perfectly by the rich and attractive MacBrayne colours of red, cream, black and gold. Following her transformation, the *Grenadier* spent the summer of 1903 on the Clyde augmenting the existing service given by the *Iona* and *Columba*, but this innovation was not repeated in later seasons. Leaving Rothesay at 8.15 every morning, the *Grenadier* sailed directly to Glasgow, returning at 1.30 for Rothesay, while on Saturdays she continued round Bute, calling at Tighnabruaich.

Contraction of Clyde services was far from the thoughts of the Caledonian directors during 1903. The withdrawal of the *Meg Merrilies* meant that a replacement steamer had to be built not only to take up her services but also to anticipate an increase in traffic from Wemyss Bay. For years after purchasing the Wemyss Bay Railway, the Caledonian had operated this single line route under considerable difficulties during the summer months; not only had all trains to slow down or stop at passing places, but the line itself was steeply graded and taxed the small Drummond 0-6-0 and 4-4-0 locomotives to their limits at peak periods. The introduction of J. F. McIntosh's '812' class 0-6-0s in 1899 eased the motive power position substantially. Fifteen of these engines, which were fitted with virtually the same type of boilers as the same designer's supremely competent *Dunalastair* express locomotives of 1896, were equipped with Westinghouse brakes for working passenger trains and many were sent to the Clyde coast section. Nevertheless, the difficulties of single line operation remained and eventually the decision was taken at the close of the nineties to double most of the Wemyss Bay line from Port Glasgow and—no less important—to rebuild and improve the station and pier at Wemyss Bay. The works were substantially forward by the summer of 1903, the first double section being brought into use on 1 June, while the new station and pier were opened on 7 December. With the latter event, the Caledonian Railway became the possessor of the finest railhead on the coast. Wemyss Bay became, and has remained, one of the showpieces of Scottish railways. The improvements involved the reclamation of a strip of land about one hundred feet wide from the shore, and upon this a new sea wall was constructed, allowing the old station to be widened. The new structure was built in Queen Anne style, half-timbered, and roughcast with red sandstone facings, and a sixty-foot clock tower dominated the new buildings. The pier was doubled in width, and thereafter could accommodate a maximum of four steamers. The passageway from pier to station was covered with a plate-glass roof. A feature of the rebuilt station was the absence of the usual private advertisements, only those of the Caledonian Railway or the Caledonian Steam Packet Company being permitted, and floral decoration was much in evidence. From being one of the worst railheads on the Clyde coast in 1900, Wemyss Bay was completely transformed by the end of 1903.

The Caledonian directors first considered a replacement vessel for the *Meg Merrilies* in September 1902, but the matter was delayed until 7 October, when a further meeting again deferred a decision. There appeared to be a disinclination on the part of some members of the board to order a new ship, and the subject kept recurring until, finally, tenders were invited towards the end of the year and on 16 December it was minuted that 'Tenders were submitted for the proposed new steamer, and that of the Fairfield Shipbuilding & Engineering Co. accepted, viz. £20,500'. The decision to go to the Govan yard for the new steamer was a fresh departure not only for the Caledonian Steam Packet Company but also for Clyde owners as a whole. The Fairfield Company had long enjoyed a well deserved reputation for deep sea and cross channel vessels, and during the nineties it had built some outstanding paddle steamers for service on the Thames, of which the *Koh-i-Noor* and *La Marguerite* were best known. Its experience of smaller paddle steamers was limited, however, and many people looked forward with interest to the company's first steamer built exclusively for Clyde passenger traffic.

Basically the new vessel was intended as a repeat of the *Duchess of Montrose*, and there were few important differences between the two. The Fairfield Company's naval architect at this time was Percy Hillhouse, who was later to become the distinguished holder of the chair of Naval Architecture at the University of Glasgow, and the new Caledonian paddle steamer was his responsibility. He produced a masterpiece. Taking the *Duchess of Montrose* as a model, he designed a steamer which was almost the equal of the *Duchess of Rothesay*, and probably the finest example ever built of a medium-speed, general purpose Clyde steamer. Her leading dimensions were virtually identical to those of the *Duchess of Montrose*, but the new steamer had greater sheer, her paddleboxes were larger, and the funnel was smartly raked. All of the aesthetic faults of the *Montrose* were eliminated, and the result was nearly perfect.

The new Caledonian steamer was launched on 9 May 1903 and was named *Duchess of Fife* in honour of H.R.H. Princess Louise, the Princess Royal. The trials were run on 5 June between the Cloch and Cumbrae lights, and the new ship delighted all concerned in her construction by achieving a mean speed of 17.55 knots, a substantial improvement on her predecessor's performance. The *Duchess of Fife* was beautifully modelled and when sailing at full speed her bow wave rose perfectly into her paddle wheels—the ideal aimed at by all designers. For a vessel intended for a speed of sixteen knots, she was brilliantly successful. The machinery was of the same triple expansion type fitted to the *Montrose*, and Navy boilers were employed.

The *Duchess of Fife* was the most typical Caledonian Clyde steamer, for she incorporated all those features which the experience of James Williamson had caused to be written into her specification—triple engines, Navy boilers, ample deck space, roomy saloons and, above all, the grace and beauty which the Glasgow public had come to expect of the Caledonian steamer at her best. She and her older consort, the *Duchess of Rothesay*, were superb, and it was no coincidence that each enjoyed a career of half a century.

Two of the older Caledonian steamers were reboilered during 1903, the *Madge Wildfire* being dealt with by the Clyde Shipbuilding & Engineering Co for £3,100, and the *Caledonia* by the same yard for £2,300. In each case the opportunity was taken of moving the navigating bridge forward of the funnel. This position became standard in the Caledonian fleet at a time when both the Glasgow & South Western and North British companies had yet to introduce it. Both vessels also received new paddle-boxes of a somewhat plain design, with horizontal slots. Other owners, too, renovated their ships during the same year, notably Captain John Williamson, who had the veteran *Benmore* overhauled and fitted wih a new haystack boiler. She was by this time the only survivor of the classic steamer of Victorian years— raised quarter deck, single engine, and low pressure haystack— all of her surviving contemporaries being of the saloon type by 1903. The Campbeltown and Glasgow Company made radical alterations to its largest steamer, the *Davaar*, which was reboilered and modernised, sailing thereafter with only one funnel instead of the two thin ones which she carried in her original state. The change was generally agreed to have enhanced the *Davaar*'s appearance.

Sunday sailing had now become respectable, and Dawson Reid continued to run the *Duchess of York*. In 1902 the Buchanans had entered the field when the *Isle of Arran* was placed on the Glasgow–Dunoon–Rothesay–Kyles of Bute run, but these two were joined in late June 1903 by none other than the first *Lord of the Isles* which Dawson Reid had retrieved from the Thames and began to operate as *Lady of the Isles* under the ownership of yet another limited company. The old *Lord* was a poor shadow of her former self and although the original colours were retained, her appearance was badly marred by ugly telescopic funnels, and those who had known her in the palmy days could be forgiven for wishing that she had gone directly to the breakers rather than revisit the scenes of her former glory. Her service was brief after her return, and she was broken up in 1904. Dawson Reid finally gave up steamer operation on her withdrawal, and the *Duchess of York* was sold to Buchanan, whose sober colours replaced the garish grey and cream of the Reid years; under her third name of *Isle of Cumbrae*, she settled down to a blameless career in the up river trade along with her old consort, the *Isle of Bute*, ex-*Guy Mannering*.

The South Western fleet received an unusual addition in 1904 as a replacement for the old *Marquis of Bute*, which was no longer adequate. It will be recalled that two identical paddle steamers were built by Russell & Co for Captain John Williamson in 1897, but that one had been sold to the south of England before completion. This vessel, which had been intended to carry the name *Kylemore* on the Clyde, but had been *Britannia* on the south coast, now returned briefly to Williamson ownership, only to be sold immediately to the Glasgow & South Western Railway to become the *Vulcan* in that fleet. The old *Marquis of Bute* was transferred to Williamson in part exchange, but went at once on charter to Ireland, being finally broken up in 1908. The *Vulcan* was not a very large steamer, being about the same size as the *Minerva*, *Glen Rosa* and *Mars*, but she was dependable and economical to run. Of modern design, fitted with compound diagonal machinery and a Navy boiler, she represented a considerable improvement on the *Marquis of Bute* and succeeded that vessel on the Millport service.

THE MARMION

The first Clyde steamer to be built for the North British Railway after the demise of the old Steam Packet Company appeared in 1906, the same season in which the other railway fleets added turbine steamers. The decision of the North British to order a paddle steamer was not dictated by conservatism, but rather by the difficulty of operating large screw vessels in the shallow water at Craigendoran and Helensburgh. The new ship replaced the *Lady Clare*, which went to Ireland, and she was built with the special requirements of the Arrochar tourist service in mind. In many respects she was a modern version of the *Lady Rowena*, and like her was lavishly equipped for tourist traffic. Of similar size to the Caledonian steamers of 1902-3, the *Marmion*, as the new vessel was called, was a smaller edition of the *Waverley*. She was fitted with a high pressure haystack boiler and compound diagonal engines. The *Marmion*'s fore saloon was extended forward of the mast and the full width of the hull, but unlike the Loch Lomond steamers and the *Lady Rowena*, the main dining accommodation was situated below the main saloon. The most important break with previous North British practice was in the placing of the bridge forward of the funnel, and the *Marmion* had the distinction of being the first Craigendoran boat so designed. It will be remembered that although the *Lady Rowena* was steered from a platform in this position, her bridge remained abaft the funnel in common with most steamers of her time.

The contract for the construction of the *Marmion* went to A. & J. Inglis, of Pointhouse, who by this period were recognised as the regular builders for the North British Company. The steamer incorporated many of the firm's characteristic features, including in particular the handsome pattern of paddlebox with eight radial vents. Her construction was well advanced by the end of April 1906 and she ran her trials on 12 June over the Gareloch measured mile, achieving 17.3 knots, no great speed by the standards of the nineties, but there was no demand for an unusually speedy vessel on the Arrochar tourist route. Rather did the *Marmion* continue the tradition of moderate speed combined with comfort which had been set by her predecessor, the *Lady Rowena*.

THE SIEGE OF MILLPORT

But the principal event of the 1906 summer was not the introduction of new railway steamers. Public attention was captured rather by what came to be known as 'the siege of Millport'. The origins of the celebrated quarrel between Millport Town Council, on the one hand, and the Caledonian and Glasgow & South

Western companies, on the other, went back to 1905. The accommodation at both Millport Old Pier and Keppel was regarded as inadequate and the Millport authorities thought it desirable to take them over and make improvements. These proposals appear to have had the blessing of the railway companies, and the town council promoted a Provisional Order in Parliament to acquire the two piers and enlarge and improve Millport Old Pier. The railways opposed the Order when it was made known that the proposed dues on vessels calling at the piers were to be as high as one shilling per ton, and a compromise was reached of only twopence per ton. After settlement of this point the Order received the Royal Assent on 4 August 1905. Millport Old Pier was thereupon purchased from the Marquis of Bute for £5,030, and Keppel from the Keppel Pier Company for £2,000, and alterations at the former were speedily taken in hand and completed by June 1906 at a cost of £9,000, including miscellaneous legal costs.

The cost of the piers and reconstruction fell on the town's ratepayers and the Council naturally felt that part should be borne by the steamer owners whom they understood to have encouraged them in the first instance. Accordingly the Council proposed that the annual compounded rate paid for steamer dues by the Caledonian Steam Packet Company, Ltd should be raised from £215 10s to £400 and, in the case of the Glasgow & South Western Railway, from £189 to £362. The increases were thought very reasonable, for the two steamer companies enjoyed important concessions, chief of which was that their vessels were not charged dues at both piers on the same run coming and going, but for one pier only. On this basis in the year 1905, the Caledonian would have been due to pay no less than £1,103 16s and the South Western £1,029 18s 6d at the twopence rate, whereas the compounded annual sum payable in lieu was already substantially lower. The Town Clerk was therefore astonished to receive a letter signed jointly by James and Alexander Williamson on behalf of their respective companies during mid-June 1906, flatly refusing to pay the increased sums. They stated that although a compromise rate of twopence per ton had been agreed, neither company had renounced the right of opposing an increase in the compounded rate. Their reason for doing so was that the service was unremunerative and could not justify the extra amount demanded by the Town Council. The latter closed with the blunt ultimatum that unless the existing rates were continued, both companies would withdraw their steamers at the end of June.

The letter caused an instant furore. The Town Council published a formal statement of its position and said that 'the entire community of Millport is in a state of great alarm and [the Council] respectfully appeal to the Board of Trade to call upon the companies for an explanation of the extraordinary threat which they have made'. Both railways, however, were adamant, and Millport readers of *The Glasgow Herald* were outraged to see a public notice on 25 June, as follows :

CALEDONIAN STEAM PACKET COMPANY
Steamboat Service with Millport and Keppel Piers
Notice is hereby given That on and from 2nd July next, and until further notice, the Steamers of the Company will cease calling at Millport and Keppel Piers.
All local and through fares and rates are hereby withdrawn and cancelled.
James Williamson, Secretary and Manager.
302, Buchanan Street, Glasgow, 23rd June, 1906.

A similar notice was inserted by the Glasgow & South Western Railway.

The steamer companies had chosen to withdraw their services at the worst possible time, just at the start of the main summer season. The basic population of the town was about 1,700 people, but the annual influx of visitors in July and August raised it to nearly 10,000. The potential loss of revenue to the citizens was therefore a serious matter, and on 26 June it was announced that uncertainty as to the outcome of the dispute had already had a disastrous effect on house letting for the two peak months. Irritated visitors rushed into print. 'A Yearly Visitor' wrote : 'Having taken apartments in Millport for July, I am anxious to

Page 197: Captains and crews: *(above)* Captain Angus Campbell of the
Columba, with the Rev Alex Stewart of Onich; *(below)* Captain A. McKellar
of the *Galatea* with crew in yacht uniforms.

Page 198 : Racing yachts : *(above)* Lord Dunraven's *Valkyrie* of 1889; *(below)* the *Valkyrie III* racing off Gourock.

know at once how to proceed in the face of the notices issued by railway companies intimating withdrawal of their steamers. . . . Are they really in earnest, and if so, do the Millport Commissioners intend to meet the difficulty in any way? Whoever is at fault, this deadlock is disastrous to everyone concerned, and the public have a right to demand that some satisfactory arrangement must be at once made. . . . Something definite must be determined on very quickly, otherwise many like myself will be compelled to cancel our engagements and find compensation from those responsible for the present position.'

'Fourteenth Summer' wrote : . . . 'it seems to me and to every regular visitor I have spoken to certain commercial suicide to the island to have this unbusinesslike muddle undecided on the very eve of the town's harvest. . . .' William Sinclair, another correspondent, expressed himself more forcefully : 'It is the duty of the Town Council to cease fooling and climb down, accept the terms offered at once, and give up their insane attempt to ruin the town and its dependent inhabitants. . . .'

Despite attempts at compromise by a committee sent to the Board of Trade in London, matters drifted during the last week of June. The weather was glorious, and the papers reported an idyllic scene at Millport. 'The sun shone from a cloudless sky. The promenade was crowded. A minstrel troupe were busy on the front, and children enjoyed themselves on the sands as only children can. Only the Magistrates and shopkeepers looked worried.' Nevertheless, the senior magistrate said that he was not in the least apprehensive of Millport being isolated, and was inclined to blame outsiders for exacerbating the situation.

An approach to the two companies by the Board of Trade was unsuccessful, neither agreeing to modify its position, and as the month drew to its close passengers booking return tickets to Millport were warned that they did so at their own risk, and that services were being withdrawn on the evening of Saturday, 30 June.

There was a last gleam of hope on 29 June when it became known that the two steamer companies had written to the Town

Council offering to agree to the *status quo* being maintained until October 1907 and in the interval to negotiate revised dues. The Council decided to call a public meeting of ratepayers and the town bellman was despatched round the streets to advertise that it would be held in the Town Hall at three o'clock that afternoon. The building held seven hundred people and it was crowded to the doors long before the meeting started. It became evident that local opinion had hardened in favour of the Council's stand, and when it was said that the railway companies had maintained that the services were unremunerative and had even offered to sell the steamers to the Burgh, those present were derisive. Two telegrams arrived during the meeting, one from Lord Bute, who had been asked to intervene but refused to do so until the railway ultimatum was withdrawn, and the second from Provost Rowatt, in London, to say that another meeting with the Board of Trade was to be held later in the afternoon. In view of this, the Millport meeting was adjourned until the evening. When the ratepayers re-assembled at eight o'clock, it was announced that all hopes of a settlement had been abandoned, and a further wire from Provost Rowatt reported the Board's encouragement in the town's resistance and suggested the proposing of a resolution condemning the railways' action. This was duly carried, and nothing remained but to see if the steamers would, in fact, be taken off.

Saturday, 30 June was a day of excitement. The fine weather continued; 'an intensely hot sun beat on the silvery sea, and holidaymakers enjoyed themselves in traditional style. The sands were alive with merry children, light-hearted youths and gaily-dressed girls promenaded along the front, and the Vaudeville company on the beach attracted hundreds.' The threat hanging over the town seemed unreal, but there was a general feeling of relief when a telegram, bearing the Broomielaw stamp, was posted up in a window of the Burgh Chamberlain's office during the morning :

Will run steamers as from Monday. Arriving afternoon to arrange details.—Buchanan.

Alas for the hopes of those on the island! The telegram was a hoax, and a disclaimer came from Glasgow at half past six in the evening:

> No telegram sent by us; know absolutely nothing of it.—Buchanan Steamers, Ltd.

The matter was reported to the police.

Large crowds gathered in the main street during the early evening and as the departure time of the last up steamers approached, there was a general move towards the pier. The *Vulcan* took the 7.10 pm run to Fairlie and her arrival was greeted by a huge crowd, amid music hall cries of 'Are we down-hearted?—No!' The steamer's deckhands removed the South Western gangways from the pier after the last passengers had gone aboard. There were cheers and counter cheers, for it was evident that the railways had their sympathisers. The *Vulcan* drew away from the pier to the accompaniment of 'Will ye no come back again?', sung by the crowd to the strains of a brass band which happened to be on the spot. As she sailed out of the bay the captain acknowledged by a long blast on the siren, and *The Glasgow Herald* recorded that 'as the good ship disappeared round the promontory on her way to Keppel Pier "Auld Lang Syne" was wafted over the waters on the wings of the evening breeze'.

The last Caledonian connection from Millport was given by the *Marchioness of Bute* which took her passengers and gangways on board and sailed off amid repetitions of the previous scenes. The crowds spent the remainder of the evening promenading along the front, pausing only to watch the *Vulcan* and *Marchioness of Breadalbane* calling at the pier on the last railway down runs to Kilchattan Bay, where they were due to spend the weekend. The *Vulcan* left almost in darkness to the strains of 'Auld Lang Syne', gracefully acknowledging by three blasts on the siren.

The Town Council had not been idle, and a public notice advertising a temporary service to the mainland at Largs via

Balloch Bay, on the east side of Cumbrae, together with services to Fairlie, was posted up on Sunday. The bellman was despatched on his rounds and it was carefully explained that this unprecedented step was being taken only as a work of 'absolute necessity'!

> Notice.—A brake will start at six o'clock prompt tomorrow for Balloch from Old Pier, Millport, thence by motor boats to Largs. Lighters and motor boats will also start at six o'clock prompt from the Old Pier for Fairlie. Connections will be kept up during the day so far as possible. . . . The conditions of this service must be accepted by passengers entirely at their own risk. Fare to Balloch, sixpence; fare to Largs, sixpence.

Four motor launches were available to carry passengers to Largs and as these were licensed for the work, no difficulty arose. It was otherwise in the case of the lighters *Craigielee* and *Elizabeth*, chartered from George Haliday, Ltd, of Rothesay, neither of which could legally trade as a passenger carrier. Various subterfuges were suggested, including the sale of postcards at sixpence each to passengers, who would technically be conveyed free of charge to Fairlie. Both vessels were chartered in the name of Councillor Cockburn, and it was asserted that passengers were sailing as his guests! The *Craigielee* took a hundred passengers to Fairlie on the dull and cheerless morning of Monday 2 July, occupying half an hour on the passage, and about 125 passengers travelled to Largs in launches. But the Board of Trade, alarmed by the inadequate life-saving equipment of the two lighters, forbade their use almost immediately on any pretext, and Millport was thrown solely on to the services of steam and motor launches to Largs. Other expedients were resorted to; amongst the more bizarre sights on the firth that week were several city men in rowing boats and lugsails.

The fun was wearing thin, however, as disillusioned travellers found to their cost when they emerged homeward bound on to the beach at Largs and were obliged to wait for lengthy periods before being picked up by launches. The Town Council wavered, and wrote to David Cooper, general manager of the Glasgow & South Western Railway, offering to settle the dispute if the railways agreed to resume their services instantly and not discontinue them again, and suggesting that an agreed annual payment should be negotiated by the parties, if necessary through the Board of Trade, such negotiation to be completed within the month. David Lloyd George, the then President of the Board of Trade, entered the dispute as a conciliator and on 4 July arranged with the three parties that the services would resume on the basis of the *status quo* until 15 May 1907, while immediate negotiations were to start with a view to agreeing upon compounded payments thereafter. With considerable relief the public learned that general agreement had been reached on these terms, and normal steamer runs resumed again on Thursday 5 July. The weather was once again 'beautifully fine'.

No reference to the coast services of 1906 would be complete without mentioning the new trains introduced by the Caledonian Railway during that summer. J. F. McIntosh, the company's locomotive, carriage and wagon superintendent, introduced a series of ten large 4-6-0 locomotives, Nos 908–917, for working the principal expresses between Glasgow and Gourock and Glasgow and Ardrossan. Based on the celebrated express engines of the *Cardean* class, built in the same year, the coast engines had the same type of large boiler, slightly shortened, and 5 ft 9 in driving wheels were used instead of 6 ft 6 in in the *Cardeans*. Two of them received the rare distinction of names, No 909 *Sir James King,* after the chairman, and No 911 *Barochan,* after the Renfrewshire home of Sir Charles Bine Renshaw, another director. These very handsome engines were the largest locomotives used by any of the three railway companies on the coast services, and along with the magnificent twelve-wheeled Clyde coast carriages, based on the Caledonian's famous *Grampian* stock for the Aberdeen route, they created something of a sensation and in publicity value alone must have done much to justify their capital cost.

The Caledonian Railway had a flair for public relations and in its beautiful locomotives, carriages and stations the Scottish public took an almost proprietorial interest and pride. The company celebrated its Diamond Jubilee in 1907 at the very peak of its achievements, certainly in so far as the Clyde steamers were concerned. The two seasons of 1906-7 marked the zenith of Clyde services in general, for after these two years gradual retrenchment began as the various owners at last undertook a rationalisation of the services. If it were possible to mark a precise date on which the decline commenced, Wednesday, 18 September 1907 would be that day. It was the occasion of one of the Clyde fleet's rare tragedies, resulting in the loss of a ship and the death of one of her crew, and from then until the outbreak of the Great War withdrawals exceeded new construction on the river.

THE LOSS OF THE KINTYRE

The morning of 18 September 1907 was very fine, with clear visibility, and as the Campbeltown Company's veteran *Kintyre* sailed down firth on her way to Campbeltown to take a special sailing to Tarbert in connection with local ram sales, her crew might have been forgiven for considering that they had not a care in the world. With no passengers or cargo, no doubt they were enjoying the sight of the firth at its incomparable best. Taking advantage of the calm sea, the new Denny-built turbine steamer *Maori*, a 3,500-ton vessel for the Union Steamship Company of New Zealand, had just completed a run at speed over the measured mile off Skelmorlie, and passed Wemyss Bay before taking a sweep prior to a second run down the mile. She and the *Kintyre* whistled, but neither slowed down, although they were on converging courses. Suddenly, and inexplicably, the two vessels were upon each other; the *Maori* failed to turn sharply enough and rammed the *Kintyre* on the starboard quarter, close to the machinery. The *Kintyre* was mortally damaged; fortunately, all but two of her crew of fifteen managed to clamber aboard the *Maori*, as the two ships remained locked together for a time, but the *Kintyre* at last broke free and sank by the stern five minutes after the collision. Captain John McKechnie, together with William Lennox, chief engineer, were seen on the bridge as she disappeared. The Caledonian steamers *Marchioness of Breadal-bane* and *Marchioness of Bute* had just left Wemyss Bay for Rothesay and Millport respectively when the accident occurred and both put back and lowered boats. Captain McKechnie was saved, but Lennox was drowned. Such was the unhappy end of the little *Kintyre*, one of the most beautifully modelled steamers ever to have sailed on the Clyde. Her loss was never made good by the Campbeltown Company, which maintained reduced services in subsequent seasons with the *Kinloch* and *Davaar*.

RAILWAY RETRENCHMENT

When the Caledonian Steam Packet Company announced at the height of the Millport dispute in 1906 that its services did not pay their way, the public was inclined to be incredulous. But such, in fact, was the case, and in the face of increasing expenses the three railway companies came together in the winter of 1907-8 and agreed to a truce in the competitive war that threatened to ruin all of them. There was a general reduction of fleets. The Glasgow & South Western Railway had disposed of the *Viceroy*, Williamson's remaining veteran, in 1907—she had been largely redundant in any case since the arrival of the turbine *Atalanta* in 1906. They followed this by selling the *Vulcan* in 1908, and a strange situation developed. She went into the fleet of Captain John Williamson, not as an additional steamer, but as a direct replacement of her sister ship, the *Strathmore*, which was sold to the Admiralty, and resumed the name which had been given her while building in 1897, *Kylemore*, but which she had never carried in service until 1908! In her new guise she became a well-known and very popular member of the Williamson fleet.

The Caledonian Steam Packet Company also disposed of a steamer in 1908, the *Marchioness of Bute* being sold for service on the Firth of Tay at the beginning of July. At the same time the Arran service was rationalised on the basis of combining the South Western and Caledonian routes from Ardrossan, the sailings to be operated by each company in alternate years in connection with both railways at the two piers in Ardrossan. The first year's operation fell to the Glasgow & South Western Railway, and the new *Duchess of Argyll* was laid up in 1908. The *Ivanhoe* was also laid up about this time and did little sailing for the Caledonian until her disposal in 1911. Both south-bank companies entered into a formal pooling arrangement from 1908, agreeing to share traffic receipts and allocate certain rosters to their steamers with the object of avoiding unnecessary competition and expense. Despite natural tensions in the early years, the arrangement generally worked satisfactorily and the joint committee formed to deal with these matters remained in being throughout the remainder of the period.

The North British Railway entered into no formal arrangement, but reduced its services in 1908. This allowed the disposal of one of its steamers, and in May 1909 the *Redgauntlet* was transferred to the Firth of Forth, where she sailed under the flag of a subsidiary, the Galloway Saloon Steam Packet Company, until the outbreak of war.

The advent of the *King Edward* on the Inveraray route in 1905 caused a major decline in the affairs of the Glasgow & Inveraray Steamboat Company. The *Lord of the Isles*, although a splendid steamer, was no match for her rival in speed, and traffic fell away. When the veteran Malcolm T. Clark died, the position of the company was so serious that attempts had been made to dispose of it, without success. It was dissolved in 1912, together with the associated Lochgoil & Lochlong Steamboat Company, but in 1909 the remaining steamers, the *Lord of the Isles* and *Edinburgh Castle*, had been transferred to the ownership of a new concern, the Lochgoil & Inveraray Steamboat Company, Ltd.

BROOMIELAW REVIVAL

Against the general run of curtailment of services and withdrawal of steamers, the announcement that a new vessel was to be built for the up-river services of Buchanan Steamers, Ltd in readiness for the summer season of 1910 came as something of a surprise. Nevertheless, the arrival of the new boat heralded a minor revival in the Broomielaw sailings about this time.

Cleansed at last of the worst sewage, the Clyde at Glasgow was no longer the evil-smelling river of the eighties and private owners enjoyed an upsurge of traffic leading, in the case of the Buchanans, to an order for a new steamer. The contract was given to A. & J. Inglis, Ltd, of Pointhouse, who sub-contracted the building of the new vessel's hull to Napier & Miller, Ltd, of Old Kilpatrick. The steamer was the same length as the North British Railway's *Kenilworth* and *Talisman*, but had two feet extra beam. She was an anachronism in the mechanical sense, being fitted with the long-outdated single diagonal type of machinery employed in the two North British boats, and steam was supplied by a low pressure haystack boiler. It was the very last example of such an installation ever to be fitted into a Clyde steamer, and its adoption twelve years after being abandoned by the conservative Craigendoran owners was due undoubtedly to the desire of the Buchanans to minimise capital outlay. The new Buchanan vessel was really a modernised version of the *Kenilworth*, although her appearance was rather different from that of the North British steamer. She received the name *Eagle III* at her launch on 14 April 1910, her owners thus perpetuating a favourite name from past years, but the use of a numeral as part of the name was definitely uncommon on the Clyde.

The *Eagle III* had a rather short fore saloon but the promenade deck was carried forward to the bow, which was left open in the manner of Caledonian steamers. The bridge was placed abaft the funnel, another old-fashioned point in her design. She had prominent paddleboxes with a large number of radial vents; by contemporary standards they were cheaply, even crudely, assembled, but this was in keeping with the general cheapness of construction. As built, the *Eagle III* was fitted with the same type of deck awning-cum-liferaft which had been used on the *Isle of Arran* in 1892, a feature peculiar to these Buchanan boats.

In the black and white colours so long familiar on the upper river, she looked very smart, and her owners expressed satisfaction with their new acquisition during the formal luncheon on the occasion of a special trip on 31 May 1910. On trials she had attained 16½ knots over the measured mile but the cruise at the end of May was run from Bridge Wharf to Greenock, Dunoon and Rothesay, and on through the Kyles of Bute to Lamlash, where the invited guests spent an hour ashore before sailing back to Glasgow.

In the light of subsequent events Dr John Inglis, of A. & J. Inglis, Ltd, must have had cause to regret some of his remarks during the cruise. In proposing 'Prosperity to the *Eagle III*' he said that 'the new vessel . . . belonged to a firm which had a knowledge of their business unrivalled by any other shipowners. They were practical men . . . and they had learned . . . that a well-constructed ship was apt to cost more than one that was not quite so well constructed.'

The *Eagle III* was advertised to make her first public sailing on the 11 o'clock sailing from the Broomielaw to the head of Loch Striven on Victoria Day, 2 June 1910. The day's sail with saloon accommodation, dinner and plain tea, cost 4s 6d. A large crowd boarded her and set off for the lower firth. Their pleasure was cut short before the steamer had gone far down the river. Those on board were appalled as she suddenly listed heavily to one side and proceeded on her way with one paddle wing practically submerged. Water poured from the vents in the paddlebox while the wheel on the other side revolved almost clear of the surface. The situation was so alarming that many of the vessel's passengers abandoned her at the first port of call. The *Eagle III* continued her sail to Loch Striven, but her behaviour was too disturbing to be glossed over. With a crowd on board she always took up a heavy list and after 25 July she was not advertised to sail again, but lay instead in Bowling Harbour while the builders discussed means of avoiding future alarms.

The *Eagle III* was taken to Pointhouse during the following winter and there, under the supervision of Dr Inglis—doubtless recalling somewhat ruefully his May speech—was the subject of a remarkable reconstruction. The steamer was in effect virtually rebuilt about herself. The trouble lay in her unduly fine under-body, and the whole of this part of the hull was rebuilt to broader dimensions forward and aft of the machinery and boiler space, increasing her displacement and altering the centre of buoyancy. The method adopted was to unplate the hull and extract every alternate frame, leaving the others to support the decks in the meantime. New frames were then put in and firmly fixed in place, after which the other frames were in turn replaced. When this process had been completed the altered hull was replated. The results of this complicated reconstruction were completely success-ful. The *Eagle III* was re-launched in the middle of March 1911 and it was seen that her stability was beyond reproach. As improved, she took up the eleven o'clock sailing from Glasgow to Rothesay in summer 1911 and the public soon forgot the antics of her first season.

The Caledonian fleet was further reduced in 1911 by the sale of two steamers. On 24 January an offer to purchase the *Ivanhoe* for £2,000 was declined, but after negotiations an increased offer of £4,000 was accepted on 7 February. The old ship took on a fresh lease of life, being sent up-river for the Broomielaw excursion trade under the flag of a new concern, the Firth of Clyde Steam Packet Co, Ltd whose title was practically the same as that of the former temperance syndicate for which the steamer had been built in 1880. She sailed at first for her new owners with all-white funnels but later these had very narrow black tops and the effect was clumsy. Two months after her disposal the Cale-donian Steam Packet Company finally bade farewell to the sturdy little *Madge Wildfire*. She too went up-river but in the ownership of Captain A. W. Cameron, of Dumbarton, a new-comer to the firth services. He purchased the vessel for £3,500 and ran her on a variety of cruises from the Broomielaw. It is understood that Captain Cameron, a 'deep sea' man, had retired early in life and acquired the *Madge Wildfire* at least partly to maintain a practical interest in shipping.

The revival of excursion traffic from Glasgow resulted in the ordering of yet another steamer in 1912, this time by Captain John Williamson. The new ship was the last to be placed in service on the Clyde before the outbreak of war in 1914. She was a paddle steamer of almost exactly the same dimensions as the *Duchess of Fife* and in many respects was a modified version of that very successful vessel. She was built by Murdoch & Murray, Ltd of Port Glasgow, and launched on 20 April 1912, being named *Queen Empress*. Her compound diagonal machinery and Navy boilers were installed by Rankin & Blackmore, of Greenock. Unlike her prototype, the *Queen Empress* was plated up at the bow and this and her plain Williamson colours gave her an austere appearance lacking in the Caledonian steamer. Her paddleboxes were attractive, and bore as centrepieces an eagle with outstretched wings, perhaps as a counterblast to the militant bird which graced the paddleboxes of the rival *Eagle III*.

Yet another steamer appeared at the Broomielaw in 1912. This was none other than the former North British Arrochar vessel, *Lady Rowena*, which returned to her old haunts under the ownership of Captain Cameron, who ran her as consort to the *Madge Wildfire*. After five years in Italian waters, she had spent another three on the south coast of England before coming back to her native firth. Her arrival was counterbalanced by the disposal of the *Isle of Bute*, ex-*Guy Mannering*, from the Buchanan fleet in July 1912 to English owners.

The Lochgoil & Inveraray Steamboat Co, Ltd gave up business in May 1912 and its two steamers, the *Lord of the Isles* and *Edinburgh Castle*, were bought by Turbine Steamers, Ltd. Fortunately their attractive colours were retained by the new owners although the *Edinburgh Castle* lasted only two more seasons before being withdrawn for scrapping at the end of 1913. The *Lord of the Isles* was taken off the Inveraray station and sailed thenceforth from Glasgow on a daily cruise round the island of Bute, on which she remained for the rest of the pre-war period.

The sale of the *Isle of Bute* was made good by Buchanan Steamers, Ltd in March, 1913 when Captain Cameron's *Madge*

Wildfire was bought in to replace her, and renamed *Isle of Skye*. Just over a year later, in May 1914, the Firth of Clyde Steam Packet Co, Ltd gave up business and sold the *Ivanhoe* to Turbine Steamers, Ltd for use on the Lochgoilhead station in succession to the *Edinburgh Castle*. The thin black tops on her funnels were thereafter deepened to the usual size.

The summer of 1914 saw the revival in the up river services in full swing, with well over a dozen steamers sailing regularly from Glasgow to the coast. Something of the colour and interest of the eighties must have returned for a short time as all these ships left the city daily for the coast resorts, even though the emphasis was now on summer excursion traffic rather than on all-the-year-round services.

After the purification schemes of the 1890s and 1900s it was possible to enjoy a trip down the river unassailed by offensive smells of sewage. Many thousands found pleasure in sailing past the Clyde's shipyards, admiring the new 'Dreadnoughts' under construction, visible symbols of Britain's international power. Few anticipated the impending catastrophe of world war and the ruin of much that seemed permanent and familiar. Sir Edward Grey's often-quoted words, 'The lamps are going out all over Europe; they will not be lit again in our lifetime', were also symbolic. The Kaiser's attack on Belgium in the first days of August 1914 heralded the end not only of many of the favourite Clyde pleasure steamers, but also of a way of life and, indeed, a generation which had brought them into being.

CHAPTER EIGHT

A CLYDE MISCELLANY

RACING RISKS

THE extent to which Clyde steamers were used before the outbreak of 'the Kaiser's War' can hardly be appreciated by present-day Glaswegians, who have largely abandoned them and favour 'stronds afar remote' at holiday seasons in preference to the shores of their own firth. Not so the citizens of the nineteenth century. All Glasgow, it seemed, went 'coasting' during the summer. Accommodation was at a premium in the Clyde resorts for speculative building never flourished in the islands of Bute and Arran, nor on the Cowal coast, and increasing demand met with no corresponding increase in housing. It was common practice for the well-to-do to take a house at the coast for the duration of the summer, from which Papa travelled daily to business in Glasgow, using the fast morning and afternoon rail and steamer services, which were on an unbelievably lavish scale by the standards of today. 'In all, 11 trains and 13 boats in connection, run for the accommodation of passengers leaving Glasgow in the half hour after 4 o'clock. On Saturdays the whole of this elaborate mechanism begins to work some two hours earlier. Needless to say, it has also to be set in motion every morning to get people up to business by half past nine; while on Mondays, in particular, the crush is so tremendous that a special relief train has to be organised in front of the ordinary daily service'—Thus Acworth, writing in 1890.[1]

It was a lucrative traffic, which encouraged such phenomena as sixty minutes for the combined rail and steamer journey from Glasgow to Rothesay via Wemyss Bay, and the even more aston-

[1] *The Railways of Scotland.*

ishing eighty minutes from Glasgow to Brodick or Whiting Bay. Stimulated by competition, the rival companies fought for the burgeoning traffic and it must be admitted that risks were taken which nowadays would be refused. The descent from Inverkip to Wemyss Bay on the Caledonian, or from the Renfrewshire hills to Prince's Pier on the Glasgow & South Western, could be hair-raising. The relatively small locomotives used during the nineties—usually 0-6-0s and 4-4-0s of moderate power—were regularly worked to their limits and it is a matter for regret that hardly any proper records have survived of their performances in the halcyon years. Considering the nature of the permanent way of those days, however, it seems certain that safety limits were exceeded from time to time on curving sections of line and it was fortunate that no serious derailment ever took place.

An intensely partisan public entered exuberantly into the spirit of the competition and rivalry amongst supporters of the various companies was fierce. Nevertheless, there were not wanting more timorous spirits whose protests, denouncing racing and calling for moderation, enlivened the columns of local newspapers even at the height of the contest. Their fears had substance, for only by good fortune were accidents avoided, or injuries prevented, on several alarming occasions. Captain Robert Morrison, of the *Duchess of Hamilton*, was involved in incidents during his first two seasons on the Arran run, due to racing against Buchanan's *Scotia*, sailing in connection with the Glasgow & South Western Railway. One of these, although humorous, could nevertheless have led to serious consequences. On 7 July 1890 the *Duchess of Hamilton* started away from Brodick pier while passengers were still embarking, and steamed off, leaving the gangway, and passengers, projecting over the edge. Insult was added to injury when the stranded victims continued their journey in the older and slower *Scotia* and were required to pay over again for the privilege, interchangeability of tickets not then having been agreed upon between the two railways.

Collisions were more spectacular occurrences in the racing days and the Arran route in particular simply invited trouble, since the competing steamers sailed from piers only fifty yards apart in Ardrossan Harbour, and access to both was by the same narrow channel. The collision between the *Scotia* and the *Duchess of Hamilton* in 1890 has already been described, but a somewhat similar episode occurred on 1 June 1899, the vessels involved being the *Marchioness of Lorne* and the *Glen Sannox*. In circumstances almost identical to those of the earlier incident the roles were reversed, the Caledonian steamer on this occasion having left for Ardrossan some twelve minutes before the Glasgow & South Western flyer followed her out of Brodick. The faster vessel had again caught up by the time the two approached Ardrossan but in this case the *Marchioness of Lorne* was deliberately steered to prevent the *Glen Sannox* from overtaking and eventually a collision took place in which the Caledonian boat was considerably damaged. In the Board of Trade inquiry which investigated the affair Captain Colin McGregor, of the *Glen Sannox*, and Eben. R. McMillan, temporary master of the *Marchioness of Lorne*, were both censured, but their certificates were not suspended, the court contenting itself with issuing a strong warning against the practice of racing.

If the Clyde steamers were fortunate in avoiding serious accidents by collision with each other, nevertheless it was too much to hope that in a period of twenty-five years there would be no fatalities from other causes. Every now and again there occurred a tragedy which cast a gloom over one or other of the holiday resorts. Often these were the result of a small boat being run down by a river steamer or swamped by its wash. A particularly sad instance of this kind of accident happened at Kirn on 1 July 1893 when a young couple with their baby were rowing in a boat close to the pier. The surge from the *Iona* capsized the boat and although all three were rescued by the steamer's crew, the child was drowned.

From time to time people fell from steamers and while many were saved there was an occasional fatality. Thus, a boy fell overboard from the *Eagle* on 1 October 1891 and was drowned, and there was some criticism of the slowness with which a boat

Page 215: Yachting steamers of 1901: (*above*) MacBrayne's veteran, the *Iona*, in Rothesay Bay; (*below*) P. & A. Campbell's Clyde-built flagship, the *Britannia*, which recalled older days by her 1901 visit, and is here seen on trials on the firth.

Page 216: Paddleboxes: *Talisman* (1896); *Duchess of Fife* (1903); *Queen Empress* (1912).

had been lowered. The North British steamers and the *King Edward* were fitted with galvanized wire mesh on their rails, which served the doubly useful purpose of preventing not only young children from falling through but also the loss of hats and other articles blown away by the wind, but this excellent feature never became general despite its obvious advantages.

There was an exciting incident at Rothesay on the evening of 20 July 1893. Just as the Glasgow & South Western steamer *Mercury* was about to leave the pier a middle-aged man, with a two-years old child, fell into the water. *The Glasgow Herald* reported that 'a young man named Joseph Maguire . . . immediately leaped from the pier to the paddlebox and seized the man, who had been stunned by a blow on the head as he fell. On being assisted by some others to pull the man out, he immediately dived after the child, who had by this time disappeared under the water. In a few seconds he came to the surface, bringing with him the child alive, and little the worse for its immersion, amid the ringing cheers of thousands on the pier and steamer. A substantial subscription was collected from the passengers by a gentleman on board, and presented to the rescuer by the captain. Many on board expressed the opinion that Maguire's bravery deserved to be brought under the notice of the Humane Society.'

THE CLYDE SEWAGE PROBLEM

The high peak of pre-war seasons came in July. The Glasgow Fair Holiday in the middle of the month saw a general cessation of work and the whole city, or so it appeared, set off for the coast. Constantly, the newspapers refer to coast trains not simply being duplicated, but being run in three, four, or even five portions to cope with the traffic. Despite the eclipse of the private steamer owners, thousands still preferred to sail from the Broomielaw which, at such times, regained something of its old popularity. But the immediate success of the new railway steamers in the early nineties was due in large measure to the fact that by travelling by rail to Gourock, Craigendoran, or Prince's Pier, the passenger could avoid the ordeal of sailing down the upper reaches of the Clyde which were then in a deplorable state of pollution due to the sewage problem. Industrial and domestic effluents were discharged into the river in substantially untreated condition, with revolting results. The smell could be appalling in a long spell of dry weather; doubtless most of the citizens were inured to it, but not so 'An Englishwoman', whose protest appeared in *The Glasgow Herald* on 3 July 1889. 'Having occasion last Saturday to enter your city by train,' she wrote, 'and being delayed for a few minutes on the bridge which spans the Clyde, I was intensely disgusted at the horrid smell which arose from the river and filled the compartment in which I was sitting. Afterwards, on walking down Jamaica Street, I noticed the same sickening stench, which appeared to reach almost if not altogether into Argyll Street. Surely the people of Glasgow must be lost to all sense of danger when they allow such a dreadful state of things to exist in the midst of their otherwise splendid city. When some terrible epidemic has broken out they will perhaps awake out of their lethargy and bemoan their folly.'

This correspondent hardly overstated the case. The simple fact was that the city of Glasgow, surging to world importance and vastly increasing in area and population during the industrial revolution, had grossly overtaxed its inadequate sewage provisions. The threat of epidemic, to which 'An Englishwoman' referred, was ever present. Tuberculosis was rife, although this disease, which was rampant until comparatively recent years, was attributable not so much to a lack of sanitary conveniences and inadequate sewers as to gross overcrowding and a lack of fresh air. But cholera and typhoid were recurrent menaces; the author's great-grandfather, who was medical officer of the then Burgh of Partick during the latter half of the century, issued official leaflets during the seventies advising as to the treatment and prevention of these diseases and no doubt his colleagues elsewhere did likewise. By the start of our period the problem had been recognised at last and a remedy sought in the gradual but thorough improvement of the condition of the river by means of the great purification schemes of the final decade of the century. But the work took time, and it is small wonder that most people preferred to use the railheads down river rather than suffer the discomfort of an up river sail. Another newspaper correspondent of the period was 'certain that Mr MacBrayne loses a large sum annually by the almost universal exodus of passengers from the *Columba* and *Iona* at Prince's Pier. . . .' By the middle nineties the problem was still bad enough to warrant a leader in the *Herald*—'For many years the insanitary condition of the river has been a public scandal. Its purification is a matter that concerns not the city alone, but the whole of the suburbs which pour their sewage into its channel. . . .'

Nor was the problem confined solely to the upper parts of the river. There were also complaints of pollution in Loch Long and Loch Goil due to dumping of sewage from the Glasgow area in these waters, and the matter was remedied only in later years when properly treated sewage was deposited instead in the outer waters of the firth off Garroch Head, Bute. Slowly things improved so that by the turn of the century there was noticeable progress, and by 1914 the up river traffic, particularly of an excursion type, began to recover. However, in 1901 a foreign visitor could still write in the following terms to *The Glasgow Herald*: 'Sir—In company with many thousands I have visited your beautiful city and Exhibition, and, with the many, have enjoyed myself much; but why does the river smell so unsavoury when one goes for a steam trip? It is obnoxious; and when I asked a lady friend if she came on the boat for fresh air, she said "She thought not! . . ." '

THE SMOKE FIEND

Possibly the only issue that raised the ire of the travelling public to the same extent as river pollution was the corresponding menace of atmospheric pollution by the steamers, a problem usually referred to as 'The Smoke Fiend'. Letter after letter, year after year, appeared in the correspondence columns of local papers, attacking the owners of river steamers, demanding legislation, suggesting remedies, and all to little avail. Thus 'Clutha', writing in 1890 : 'I have been taking advantage of this fair May time to take an outing "doon the water", and wherever I go I am disgusted to find the beauty of the scene and the purity of the air befouled by long impenetrable banks of the densest smoke, vomited forth by steamer after steamer. . . . On some days recently, when the air was heavy, a dense pall of blackness has hung undispelled the whole day over the whole space between Greenock, Dunoon, Rothesay, and the Cumbraes. . . .' In the same year another writer said : 'I am glad to see public attention drawn to the river smoke evil. . . . I live at Gourock, having fled from Glasgow smoke; but the fiend has come hither by river steamer.' In 1891, 'No smoke' wrote to say that 'last Thursday I left Gourock at 5 pm; there were four Caledonian steamers leaving and two South-Western steamers passing at the time, but where they were no one could tell; all was smoke ! smoke ! . . .'

The smoke from steamers was sufficiently heavy to cause pollution of the waters of the firth itself, as Alfred E. Fletcher, H.M. Inspector under the Rivers Pollution Prevention Act reported in February 1888 after an examination of Loch Long and Loch Goil. 'The soot caused by the numerous steamboats plying on the lochs and on the Clyde falls on the water, and in calm weather collects in patches. It may be observed floating with the tide and drifting into the lochs. The whole of this, and the black smoke sent forth from the steamboats' funnels, is curable, and that without loss to those who now carelessly produce it. Pressure is needed to compel the constructors of the boilers and those who use them, to cause the nuisance to cease. At present the practice is not forbidden, except when the vessel is alongside a pier or wharf. . . .'

Fletcher's comment on smoke emission while steamers were at piers was of little comfort to passengers. Certainly it allowed local authorities to prosecute offenders and even such august skippers as Captain Angus Campbell, of the *Columba* and Captain Donald Downie, of the *Lord of the Isles,* were successfully dealt with at Rothesay in the summer of 1891 for offences against the regulations. But once a steamer was on the move, nothing could be done. Not only was the smoke in itself a nuisance, but it also carried hot cinders, especially under hard steaming conditions. Many photographs of this period show passengers holding raised parasols or umbrellas in obviously fine weather to protect clothing from damage.

There was little improvement. Steamer owners were not inclined to fit more expensive boilers simply to reduce smoke, and with abundant supplies of good quality coal cheaply available from the Lanarkshire coalfields there was limited incentive to do so to effect marginal economies either. The haystack boiler was cheap to construct and install and had many advantages, so why change to something more elaborate? There had been an experiment with coal briquettes on the *Edinburgh Castle* of the Lochgoil Company in 1888, but it came to nothing although it was stated that smoke had been somewhat reduced. The Caledonian Steam Packet Company conducted a bolder experiment in 1893 when the *Caledonia* was fitted for the burning of oil fuel instead of coal. This gave very good results, and the company's enterprise was favourably received by the public. 'I was much struck by the entire absence of smoke from the funnel,' wrote a passenger; 'I learnt on inquiry that this was owing to the burning of oil instead of coal. . . . The Caledonian Company . . . is surely worthy of imitation. . . .' Another wrote in similar terms : 'I cannot refrain from complimenting the Caledonian Steam Packet Company on their enterprise in introducing oil fuel and bringing their experiments to so successful an issue, and think that every credit is due to the company for their success in expelling in some measure our old enemy, the smoke fiend. . . . Before travelling by the Caledonian I must confess to being prejudiced against oil fuel, but anyone who has seen the results of burning oil cannot for a moment doubt the efficiency of the process, as the spotless decks and cleanliness of the boat throughout testify to the fact of an absence of soot or dirt, which is usually so much in evidence on pleasure steamers.'

Captain James Williamson himself wrote that the six-months'

test had proved successful and had fully established the superiority of oil over coal. Quite apart from the alleviation of the smoke nuisance, it was noted that the *Caledonia*'s speed was higher with oil burning. Unfortunately, the price of oil was prohibitive and regretfully the system was abandoned. It was revived in 1897 in the form of a combined coal- and oil-burning system by Farmer & Stewart, of Glasgow, and again, although successful, it was given up in due course. The next attempt to control smoke came in the turbine steamers of the Edwardian period. The *Queen Alexandra* of 1902 was the first to be fitted with the Denny & Brace spark arrester which did much to stop showers of soot and cinders falling on the decks. Nevertheless, despite these attempts, there was never a general effort to eliminate the nuisance caused by smoke. Even *Punch* published a cartoon showing the beauties of the Clyde almost completely obliterated by vast palls of smoke, which many travellers agreed was hardly an exaggeration. By the general public, however, the problem was accepted with resignation as an unavoidable concomitant of travel by steamer and those who protested fought, on the whole, a losing battle.

FOG AND STORMS

Bad weather on the firth was another hazard which assailed travellers by steamer from time to time, although the fairly sheltered waters of the Clyde estuary usually allowed services to be maintained except in the worst conditions. Fog was the danger most feared and scarcely a winter passed without some delay or stoppage of the boats for this reason. Usually the Broomielaw steamers were affected before any of the others, due to the tendency of the fog to gather near Glasgow. Some of the worst weather during the whole of our period was experienced in the winter of 1894-5—the year of 'the great frost'—and on 8 February a dense fog caused the *Benmore* to tie up at Bowling in the course of her up run from Rothesay, and she had to remain for three hours before it was safe to proceed. This vessel was the victim of what was probably the worst fog delay ever recorded;

about 1910, she was caught off Dumbarton and had to anchor for *two days* before it was possible to proceed to Glasgow. There was understandable concern for her safety, as it was impossible to find out what had happened until the fog lifted.

The 1895 winter continued to be punctuated by foggy weather and on 12 February it was reported that there was a dense fog in Glasgow and in consequence traffic on the river was almost completely suspended; the *Kintyre* managed to get away on her run to Campbeltown, but arrived at Gourock no less than five hours late. On 1 March there was more fog and the *Edinburgh Castle* was unable to leave Glasgow for Lochgoilhead, so that the *Benmore* and *Meg Merrilies* deputed for her between Gourock and Lochgoilhead, both vessels having themselves been unable to sail to Glasgow in view of the conditions.

Storms were more common than fog on the outer firth and often caused delay due to difficulty in taking the piers. A particularly bad spell of such weather began on 13 October 1891 when a heavy sea running off Gourock Pier gave trouble to the skippers of steamers calling there. On the next day the storm badly damaged the Dunoon Pier, causing much inconvenience. There was a lull of a few days, but on 19 October the pier suffered further damage in a second storm. Usually damage of this kind was rare and apart from the inevitable repercussions on services nobody suffered unduly, unless from seasickness. But on 3 December 1891 there was a tragic exception when a violent storm burst over the Skelmorlie district. The *Caledonia* had much trouble in coming alongside the pier at Wemyss Bay, but eventually did so and disembarked her passengers. However, as she left the pier Muir, the mate, was washed overboard and drowned. A boat was lowered, manned by two passengers, but they were unable to save Muir and they themselves reached the shore only with considerable effort and at great personal risk.

Now and again rough weather even endangered vessels in harbour, as for example when a north-easterly gale blew up in the early hours of Saturday, 31 May 1902 and reached hurricane force about three o'clock in the morning. Masters of steamers in Rothesay Bay had to raise steam in case the moorings gave way and the ships were blown ashore. The *Mercury*, due to leave at 7.45 am, had her bows badly smashed when a mooring rope snapped, and the *Iona* lost twenty feet of her stern rail when leaving the pier later in the day.

Perhaps the most vivid description of one of the Clyde paddle steamers in a storm appeared in *The Glasgow Herald* of 17 November 1888, which told of a gale on the previous day, and it is worth quoting as an example of what the vessels on the winter services were sometimes called upon to face :

> On the wide stretch of water between Greenock and Helensburgh, and throughout the entire firth, the river was in a perfect boil which no vessel would care to encounter. . . . The *Grenadier* had an exceedingly stiff time of it in her attempt to proceed on her way to Ardrishaig. The captain, apparently, did not care about giving up, and he bravely piloted his vessel across the firth to the Argyll shore. On rounding Battery Point and Kempock the seas were running very high, and the run was made in the midst of much danger. In getting near to the shore it was found impossible to get close to Kirn or Dunoon piers, the water lashing over them with great fury. Still the skipper did not give up hope of getting along, and he then steered for Innellan, keeping well out. Coming opposite Innellan, he saw that his attempt to reach it would be an impossibility, and he also recognised the fact that there was no hope of rounding Toward. By this time the waves were raging over the steamer, getting as high as the bridge. The people on board gave themselves up as lost. The deck was cleared of everything movable, and all who had no business above were sent below out of the way. The run up to Greenock was undertaken with much greater speed, the wind being astern, but still the waves lashed over the boat. Prince's Pier was reached in safety, everybody thankful that their terrible experience was over.

YACHTING

The tribulations of winter were easily forgotten, however, in the height of summer, when the Clyde was at its gayest and all of the excursion steamers, 'the summer butterflies', were in service. Probably the high point of the year was the Clyde Fortnight, the annual two weeks of yacht racing in the first half of July organised by the Clyde yacht clubs. This was the era of the legendary racing cutters designed by William Fife, of Fairlie, and the great George Lennox Watson. The Prince of Wales himself ordered his superb yacht *Britannia* from D. & W. Henderson, of Meadowside, Partick, in 1893 and alongside her was built the *Valkyrie II* for the unlucky Lord Dunraven, both to Watson's designs. Yacht racing became a popular Clyde entertainment. It was customary for one or more of the Clyde steamers to be chartered as 'Club Steamer' during the Fortnight, and the Caledonian *Duchess of Hamilton* came to be associated more closely than any other vessel with these duties. In earlier seasons, however, the *Victoria* and *Galatea* were often employed, as was the new *Talisman* in 1896. The Caledonian Steam Packet Company's crews wore white Navy jerseys when on yacht duty. The club steamer followed the yachts round the racing courses, carrying officials, guests, and friends of the members, and more often than not a military band or light orchestra was engaged for the occasion. When the *Galatea* was club steamer in 1891 she carried 'a military band . . . under the direction of Herr Iff', and the *Victoria* a year later conveyed the band of the 1st Volunteer Battalion, Highland Light Infantry.

The extent of the racing in the nineties is indicated by the fact that some seventy yachts were under canvas on the occasion of the Royal Clyde Yacht Club's Regatta on 11 July 1891. What a period picture is conjured up by the newspaper report! 'The steamer "Galatea" was the club steamer, and with the crowd on board just escaped from violating her certificate. The parasols made the most brilliant flower show of the season, a compliment no doubt intended for the member in charge. . . .'

Interest was not confined to yacht club members. The steamer owners were quick to recognise a source of revenue and pleasure trips were arranged every year to allow the public at large to follow the fortunes of the great racing cutters round the various courses. Now and then the prospect of a financial windfall tempted the operators into chartering inferior vessels for the ordinary service runs, so as to release the steamers for yachting excursions. The Caledonian Steam Packet Company's arrangements for maintaining the Millport and Kilchattan Bay service on 7 July 1894 drew a furious letter of protest from a Mr Thomson, of Kilchattan Bay. 'Do the Caledonian Company think that the visitors to Kilchattan are a lot of tramps?' demanded the irate correspondent. 'Coming down last Saturday per 2.40 pm train to Wemyss Bay, the passengers to Kilchattan were bundled into a dirty tug-boat which was unfit for swine. I do not intend ever coming again per Wemyss Bay, neither do the rest of my fellow-travellers. There is a good connection via Fairlie, where the passengers are never so treated.'

This was the year in which the Clyde Fortnight was marred by the accidental sinking of Lord Dunraven's *Valkyrie II*, the unsuccessful challenger for the *America*'s Cup in 1893. While sailing in the Holy Loch, she was in collision with the *Satanita* and sank shortly afterwards in deep water, one of her crew unhappily receiving fatal injuries. She was replaced in 1895 by the *Valkyrie III*, built by D. & W. Henderson, which contended for the *America*'s Cup in that year but her challenge was one which, for various reasons, led to much ill-feeling between British and American yachtsmen and it was for the popular Sir Thomas Lipton to restore good relations by challenging anew with his successive *Shamrocks* from 1899 onwards.

Public interest in the large yachts was immense during the nineties and it was common for crowds to gather at strategic points on shore to watch them racing. During the Royal Clyde Yacht Club's Regatta on 7 July 1895, when the *Valkyrie III* won her race against *Ailsa* and *Britannia*, *The Glasgow Herald* reported that 'the number of spectators at the regatta was beyond calculation. From Hunter's Quay along to Dunoon and on a distance towards Innellan the shore was black with people; on the other side of the firth the whole line of coast held many thousands of onlookers; in the neighbourhood of Kilcreggan it was much the same; and the club steamer was literally packed!'

A CAMPBELL VISITOR

Perhaps the most interesting year of all as it affected the Clyde steamers was 1901, the year of the Glasgow Exhibition. A special series of International Races was run in June under the auspices of the Exhibition authorities, and several river steamers were used to follow the yachts. Not only were the main Clyde fleets represented but Peter and Alexander Campbell thought the occasion sufficiently attractive to send their flagship, the *Britannia*, from Bristol to take part as well. Here was a fascinating, although brief, episode recalling for many people the great days of the Kilmun fleet when the all-white funnel was a commonplace sight on the firth. The Campbells never forgot their Scottish connections, and made a practice of ordering all their new ships from Scottish yards. The *Britannia* had been built by S. McKnight & Co, of Ayr, in 1896 and she was regularly commanded by Captain Peter Campbell himself. Her Clyde excursion of 1901 was advertised in the *Western Mail*:

Glasgow International Exhibition and Yacht Races. Special Long Trip from Cardiff, Penarth and Mumbles to the Clyde, by the magnificent seagoing Saloon Steamer 'Britannia'.

(Weather and circumstances permitting.)

Thursday June 6th, 1901. Leave Bristol (Hotwells) 9.15 a.m. Cardiff 11 a.m. Penarth 11.10 a.m. Mumbles 1 p.m. thence proceeding direct to the Clyde, arriving at Gourock early the following morning.

On Friday and Saturday, June 7th and 8th, the steamer will leave Gourock (after the arrival of the 8.30 a.m. train from Central Station, Glasgow) for Rothesay from which place she will follow the yachts, returning after the races each day to Gourock (from the latter place trains run every hour to Glasgow).

Tuesday June 11th, the steamer will leave Gourock after the arrival of the 8.30 a.m. train from Central Station, Glasgow, and will proceed round the Gareloch, Loch Long, Holy Loch, passing Dunoon, Toward Point, across Rothesay Bay, through the Kyles of Bute passing close to the Island of Arran, giving a splendid view of the far-famed Goat Fell, Corry, Brodick, Lamlash, Holy-

Isle and Ailsa Craig, arriving in the Bristol Channel on Wednesday June 12th as follows, Mumbles 9.30 a.m., Cardiff 12 noon, Bristol 2 p.m. Return Fare £2.10/- 10/- extra to follow the Races. Number limited to 200 passengers, for whom sleeping accommodation will be provided while on the steamer going to and returning from the Clyde. Berths will be allotted in rotation and intending passengers are required to make early application for tickets.

The races took place on Friday and Saturday, 7 and 8 June, in perfect weather—too fine, really, for the yachtsmen, although the *Herald* thought that 'it will be generally agreed that from the mere pageantry point of view the gathering was the source of much public gratification'. Surely few other occasions in Clyde steamer history can have brought together such a colourful assembly of famous paddle steamers; as well as the *Britannia*, the North British Steam Packet Company's *Waverley*, the Glasgow & South Western *Juno*, and Buchanan's *Isle of Arran* were advertised to follow the yachts on both days, while on the Saturday the *Isle of Bute* and *Duchess of York* reinforced the fleet, together with the *Iona*, which carried an official party. There is, unfortunately, no record of which steamers were chartered by the yacht clubs for the benefit of their own members, but it is not unreasonable to assume that the Caledonian Steam Packet Company contributed at least one vessel on this important occasion. The first race on 8 June was over a course of 47 miles, starting at Rothesay, the prize value being £250, and the yachts which competed were Sir Thomas Lipton's *Shamrock*, Whitaker Wright's *Sybarita*, Kenneth Clark's *Kariad*, and the Kaiser's *Meteor*.

MUSIC ON BOARD

In the spacious years before the Great War, no pleasure trip on board a Clyde steamer was complete without the musical accompaniment which was a characteristic feature of the Clyde scene, ranging from itinerant fiddlers on the older, privately-owned up-river boats to the popular German bands of the railway steamers.

The late Cameron Somerville has drawn attention to the difference between the programmes offered by the groups on the Broomielaw vessels and those of the steam packet companies. The latter, usually Bavarians or Württembergers, offered a varied selection of light classical music, Strauss waltzes, and popular ballads, whereas the former tended to concentrate more on the native Scottish songs and, doubtless, music hall favourites of the day. Public opinion, ever quick to suspect, denounced the Germans as spies in the jingoistic atmosphere of the early months of the Great War, and the author has himself been solemnly assured of this 'fact' in the 1960s by an elderly passenger on board a river steamer! This old, but persistent, canard seems to have been supported by the presence of an Admiralty torpedo testing range in Loch Long, to which the harmless musicians are reputed to have devoted much attention.

The Caledonian Steam Packet Company had an arrangement for many years with the Hayward family, who toured the country giving concerts during the winter, but appeared on the company's steamers in summer under the name of the Brescian Family Concert Party, providing an appropriate background for specially-advertised afternoon and evening musical cruises from the principal coast resorts. But the most amazing musical occasions of all were due to the resourcefulness of Dawson Reid. Anxious to prove the respectability of his Sunday cruises, he saw his opportunity during 1901 and the astonished citizens of Glasgow read in the advertisement columns of *The Glasgow Herald* of 7 June of the proposed cruise of the *Duchess of York* to 'The Far Famed Kyles of Bute', carrying on board the Berliner Philharmonisches Blas-Orchester (presently appearing at the Exhibition) conducted by Herr Moser, who would give a 'Grand Al Fresco Sacred Concert'. One of the most astonishing spectacles of the whole period must have been that of the *Duchess of York* solemnly paddling up the Kyles of Bute to a varied accompaniment of Mendelssohn, Stainer and Sullivan!

Most of the musicians on the steamers were capable and enjoyed a well-deserved reputation, but not so some of the

INTERNATIONAL YACHT RACES.

JUNE 7TH, 1901.

SPECIAL SATURDAY AFTERNOON CRUISE—

TO DUNOON AND BACK FOR 1s 6d,

By Saloon Steamer

"DUCHESS OF YORK,"

From GLASGOW (BRIDGE WHARF)

AT 1.55 P.M.,

Calling at PARTICK, 2.10; GOVAN, 2.15; RENFREW, 2.45; DUNOON, 4.30 P.M. Cruising round the Yachts in Rothesay Bay.

RETURN FARES.

Day's Cruise—Fore Saloon, 2s; Saloon, 2s 6d.

Returning from Dunoon at 6.

Arriving in Glasgow at 8.15.

A. DAWSON REID,

53 Bothwell Street, Glasgow.

GLASGOW INTERNATIONAL EXHIBITION.

SUNDAY ON THE FIRTH OF CLYDE.

GRAND AL FRESCO SACRED CONCERT,

SUNDAY, JUNE 9TH, 1901,

BY THE BERLINER PHILHARMONISCHES BLAS-ORCHESTER

(Presently appearing at the Exhibition). Conductor—Herr MOSER.

TO DUNOON, ROTHESAY,

THE FAR FAMED KYLES OF BUTE,

Cruising round Loch Ridden and the Burnt Islands,

By the Swift Saloon Steamer

"DUCHESS OF YORK."

From Glasgow (Bridge Wharf) at 10.45 A.M.

Calling at Renfrew and Bowling. Returning from Rothesay at 5 P.M., Dunoon 5.45 P.M., arriving in Glasgow at 8.15 P.M.

RETURN FARES for This Day only.

	Fore-Saloon.		Saloon.
To Dunoon	3/6	....	4/6
„ Rothesay	4/6	....	5/6
„ Kyles of Bute	5/	....	6/6

A. DAWSON REID,

53 Bothwell Street, Glasgow.

Dawson Reid's cruise advertisements of 1901

vagrant type who were more commonly encountered in the earlier years. A letter which appeared in *The Glasgow Herald* in June 1888 speaks for itself:

I left at 10.30 and returned in time for dinner, after a sail through the most lovely scenery on our western coast, full of the reminiscences of a most enjoyable day but for one discordant feature. From the moment we left Glasgow Bridge Wharf until we returned, two youths, armed with excruciating instruments of torture, took possession of the saloon gangway. A penny brass whistle and a cracked violin—the whistle one of the shrillest I think I have ever heard, and the violin, a most *curioso* arrangement of wood and string, emitting the most terribly rasping sounds—never ceased through those long hours. Go where one would, the maddening sounds followed. In desperation, I gave the musicians a shilling to play amongst a crowd of Sunday-school children in the forecastle; but apparently they were incontinently hounded out, as they re-appeared, piping more vigorously than ever, in a few minutes. No scenery, no air, no luxury can compensate for being exposed to the torture of having to listen to the terrible musicians who apparently frequent our coast steamers.

PIER DUES

A more general cause of complaint amongst coast travellers than pennywhistles lay in the practically universal system of pier dues which obtained on the Clyde during the period. The main railway piers were free to the passenger but at most of the smaller piers on the firth and Loch Lomond he was charged varying sums for the privilege of embarking or disembarking from the steamers. Had the tolls been uniformly applied it is probable that there would have been no undue protest, but some proprietors charged heavily by the standards of the time; others, more generous in their attitude, allowed useful concessions. When Innellan Pier was enlarged and re-opened in May 1900, for example, the attention of the public was drawn to the fact that visitors having arrived at Innellan would be permitted to use the pier without further dues while their holidays lasted, a privilege extending also to their families and servants. The charge at this pier was one

penny per person, but in Arran the dues were heavier; at the start
of the period passengers were charged 2d each way at Brodick
and Lamlash, and no less than 4d at Lochranza, while at Pirnmill Ferry each passenger was charged 6d, no small sum at the
time. Some hard things were said by aggrieved travellers, some
of whose protests in the columns of the press verged on defama-
tion. Writing in the autumn of 1888, a correspondent threw the
blame 'on the natives of Lochranza themselves, who, *with that
greed so inseparable from the Arran character,*[1] have tried to
extort as much as they possibly could from those whom they consider their natural prey—the summer visitors. . . . I think the
attention of the Duke [of Hamilton, the proprietor of Arran]
should be drawn to the shameless extortion going on at Lochranza. . . .' Three years later, another traveller, writing to complain of the dangers of the ferry system at Whiting Bay (the pier
had not then been built) took the opportunity of a passing tilt at
'the natives'—'Last week, while strong east wind prevailed, one
of the ferry boats was nearly swamped in going out to the Glen
Sannox. Indeed it alarmed the passengers and the boatmen so
much that the ferry fares were not paid or uplifted, and Arran
visitors know what that means.'

The feud between visitors and pier owners went on for years,
with occasional notable victories, such as the abolition of dues at
Helensburgh in 1896, but one of the most amusing incidents was
recorded by a writer to the *Ardrossan and Saltcoats Herald* in
June 1890 who wrote :

> Sir—Pier dues are generally paid with a grudge, and the following
> incident probably illustrates pretty accurately the feeling of most
> people toward the modern Matthew. A fortnight ago, on the first
> Saturday afternoon excursion by the 'Duchess of Hamilton',
> several passengers landed on Lamlash Pier, and as usual had to
> pay 2d each before being allowed to pass to *terra firma,* In less
> than an hour most of them returned to the steamer, and had to
> part with other 2d before going on board. But one individual
> objected. Apparently deeming it an imposition in a double sense
> that he should be asked to pay 4d for a sojourn of 40 minutes

[1] author's italics.

on the island, he would not, as requested, stand and deliver. Then ensued a fracas. Backed up by those behind, the indignant objector strove to force his way past the collector. But tax-gatherers are not so easily got rid of. Laying hold of the bellicose disputant, the collector angrily demanded payment. His opponent only tried the harder to escape. Then they closed, and, struggling, fell, rolling over one another on the pier. This was just what the others had been hoping for, and while these two were hugging one another like a couple of amatory bears, the remaining passengers hastily made their way aboard the 'Duchess'. By-and-by the hero of the riot also arrived, having purchased his freedom by suffering himself to be mulcted in 2d under protest; and the laughing and chuckling of the fortunate ones, as they related their experience to the creator of the disturbance, almost compensated him for the rough usage he had sustained at the hands of 'the man at the quay'.

CHAPTER NINE

BOILERS AND MACHINERY

NOT the least of the attractions of a pre-1914 Clyde steamer was its engine room where, amid the intermingled aromas of hot oil, cooking, and sea air, a passenger could watch the machinery at his leisure. The marine reciprocating engine is impressive to see in action and Clyde passengers of this period were privileged in knowing some of its finest examples. Painted and decorated entablatures, bright steelwork and polished brass fittings provided the setting in which swinging connecting rods exercised their fascination on young and old alike, while on some vessels the access doors to the paddles were left open, permitting a view of the wheels in action. The turbine steamers lacked the obvious appeal of moving parts, but had a subtle attraction all their own, in utter contrast to their paddle sisters.

For such enjoyment, as for so many other things, the passenger had to thank Captain James Williamson who was primarily responsible for transforming the dingy, boxed-in machinery spaces of the older vessels into the brightly lit, smartly turned out engine-rooms which characterised the Caledonian steamers and subsequently those of other fleets. Aware of the engines' value as a spectacle to attract passengers, Captain Williamson replaced the wooden walls of the older tradition by open rails, through which even the smallest youngster could see without effort. His engineers wore smart uniforms, redolent of naval practice and, following the example of the Caledonian company, the engine-rooms of Clyde steamers acquired an enviable reputation for smartness and efficiency which has continued to the present day.

SINGLE DIAGONALS AND HAYSTACKS

We have seen that in 1889 the average Clyde paddle steamer had a fairly cheap arrangement of engine and boiler. The combination of haystack boiler and single diagonal engine was well-tried, reliable, and, above all, cheap to install. The word 'diagonal' in-

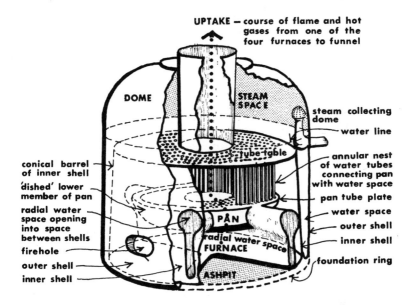

DIAGRAMMATIC SECTION OF A HAYSTACK BOILER

A typical haystack boiler

dicated a reciprocating engine differing from the well-known horizontal engine, familiar for so many years in factories, in having the cylinder placed below the level of the crankshaft and inclined 'diagonally' upwards towards it, so permitting the heavy casting to be placed as low as possible within the hull. Most vessels at the start of the period had machinery of this type, and its disadvantages did not outweigh its cheapness in the view of owners of small private fleets. This kind of machinery was, of course, the most elementary form of reciprocating engine and its thermal efficiency was low. Utilising low pressure steam from a haystack boiler, fuel consumption could be very high in large vessels of this type. Coal, however, was cheap, particularly in the eighties and early nineties. The minutes of the Caledonian Railway record purchases of coal for locomotive purposes at 3s 8d *per ton* and at such prices an owner could afford to disregard the relatively academic matter of thermal efficiency. The other disadvantage of the single engine affected passengers more directly— being an unbalanced force, the large piston sliding to and fro in the cylinder set up a pronounced and sometimes unpleasant surging motion when the vessel was sailing. Nevertheless, pier to pier runs were generally short, and the passenger seldom had to endure the sensation for long, unbroken periods.

Although elementary in principle, a single diagonal engine was massively constructed. The connecting rod and crank webs were placed within large cast iron entablatures which were bolted to the cylinder casting and embraced the slide bars. A single eccentric sufficed to operate the slide valve, since it was not the practice to work such machinery expansively as on a railway engine. Occasionally a steamer stalled on 'dead centre' and a pinch bar was used to move it; later examples of this type were fitted with an auxiliary starting engine to save resorting to such crude expedients and some of the older ships were later modified in this respect.

The haystack boiler, so often paired with the single engine, was a primitive form of water tube boiler. Externally it resembled the haystack from which it derived its nickname. There were usually four furnaces at the foot, separated from each other, and the hot gases rose round a large, water-filled pan and mingled amongst a nest of water tubes connecting it with the dome before rising into the funnel uptake. The haystack was a light boiler and was highly regarded as a rapid steam raiser but under natural draught it was not suitable for very high pressures and tended to fall out of favour with the introduction of compound machinery. Reference has been made to P. & A. Campbell's *Westward Ho!* as being the first example of a pleasure steamer of Clyde type to combine compound engines with a haystack, and the North British fleet was fitted with 'improved haystacks' right up to 1914, while MacBrayne and Buchanan were two other owners who continued to use them.

COMPOUNDS AND NAVY BOILERS

Captain James Williamson was early convinced of the advantages of compound machinery, the basic principle of which is to use steam to drive a piston in a small high pressure cylinder and afterwards to use it at a lower pressure in a larger cylinder before finally exhausting it to the condenser. More economical use is thus made of the steam, with a corresponding reduction in fuel consumption as compared with a single engine of the same power. Two cylinders, however, normally suppose an engine with two cranks—in effect, a double engine—with increased initial costs, and it was probably to keep costs down that Captain Williamson favoured the tandem arrangement of cylinders, with a high pressure and low pressure cylinder in line, the piston in each driving a common piston rod, connecting rod and crank. This simple and effective arrangement saved space and avoided the cost of twin crank machinery, albeit the surging effect of the old single engine was if anything aggravated by the addition of the extra piston in the high pressure cylinder. Nevertheless, in vessels intended for short distance work on the upper firth no doubt it was felt that the additional expense of a twin crank engine was not justified.

With compound engines came improved boilers, the most popular of which was the Navy, or locomotive type. It was a cylindrical boiler with horizontal fire tubes leading from the combustion chamber to the uptake, and a prominent feature of steamers so fitted was the placing of the funnel further forward than usual in vessels with haystack boilers, generally to the detriment of their appearance. The *Caledonia* was an example of this defect in outline, but the *Duchess of Fife* was an exception to the rule in having a well balanced profile.

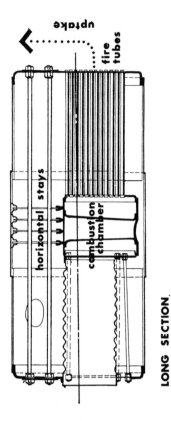

LONG SECTION.

uptake

fire tubes

horizontal stays

combustion chamber

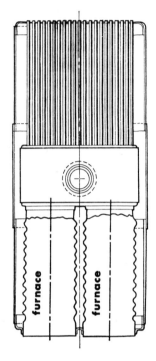

PLAN

Navy boiler

furnace

furnace

CROSS SECTION

Navy boilers operated in closed stokeholds under forced draught induced by fans. Older vessels with more primitive forms of boiler used natural draught which of course varied according to weather or, indeed, the speed of the ship, but forced draught guaranteed constant conditions. In the early days lack of experience led to excessive draught being used and a writer in one of the newspapers in 1890 complained of ash and live cinders falling on the promenade deck of the *Caledonia* because of this! These were teething troubles, however, and the Navy boilers gave generally reliable service in the Caledonian and Glasgow & South Western fleets.

When the North British Steam Packet Company had the *Meg Merrilies* built in 1884 she was fitted with a twin crank single expansion engine, an expensive installation compared with a single diagonal but vastly superior in avoiding discomfort due to fore-and-aft motion. The *Victoria*, of the Wemyss Bay fleet, was another notable example. Writing to *The Glasgow Herald* a correspondent spoke of 'the Meg's beautiful engine' and her freedom from noise and movement compared with the newer Caledonian tandem compound steamers. But the *Meg*'s form of machinery never became really popular; its sole advantage over the single diagonal lay in absence of surging and its increased initial cost without the corresponding advantages of compounding made it unpopular. The *Meg Merrilies* herself was duly converted to compounding in 1898 by replacing her two original cylinders of equal diameter by new ones, one high pressure and one low pressure, to her considerable improvement.

Once the principle of compounding had been accepted the obvious choice of machinery, for the larger steamers at any rate, was the twin crank diagonal engine. The first of this type was fitted into the *Galatea* but it was widely regarded as being too powerful for the light hull of that imposing vessel. The *Duchess of Hamilton*, a Denny steamer, had one of that firm's standard compound engines, instantly recognizable by its Brock valve gear, a variant of the Walschaerts locomotive valve gear for marine use, designed by the managing director, Walter Brock. This re-

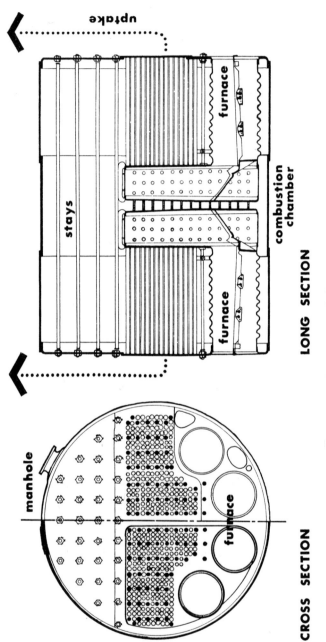

uptake

furnace

stays

combustion chamber

furnace

LONG SECTION

manhole

furnace

CROSS SECTION

Double-ended boiler

quired a very distinctive pattern of entablature, lower, and tilted further forward to support the expansion links, than in a steamer with Stephenson gear. The latter type, in one form or another, was standard on other steamers, usually with long eccentric rods and expansion links close to the cylinder casting, but the *Waverley* of 1899 had shorter rods, so that the expansion links were positioned in full view about midway between entablatures and cylinders.

In the twin crank compound diagonal engine Clyde paddle machinery reached its finest development. Each builder turned out a characteristic version of the type, of which the Denny variant was probably the most unusual, but builders such as A. & J. Inglis and J. & G. Thomson produced very fine examples. Details were refined and improved as the years passed and aesthetically many sets of machinery were delights to the eye. Amongst the finest of all, the engines of the *Duchess of Rothesay* and the *Jupiter* took a high place. Glasgow & South Western steamers had probably the most powerful machinery of this type, being built for speed regardless of expense, and the *Juno* and *Glen Sannox* had compound engines with much larger moving parts than those of most of their contemporaries, the rods and crank webs of the former vessel being particularly massive.

TRIPLE EXPANSION

In the Caledonian Steam Packet Company's triple expansion steamers *Marchioness of Lorne*, *Duchess of Montrose* and *Duchess of Fife* the tandem principle was further developed to afford economies in space and first cost as compared with a triple engine with three cylinders and cranks. The triple expansion principle is the logical development of compounding and in its simplest form steam from the boiler passes at decreasing pressures through three increasingly larger cylinders before exhausting to the condenser. This usually requires three cranks, but the Caledonian vessels had two small high pressure cylinders, one each in tandem with the intermediate and low pressure cylinders, which permitted a twin crank arrangement, economising alike in cost and space, just as the *Caledonia*'s pattern of tandem compound engine had been superior in this respect to a conventional set of compound machinery. The *Marchioness of Lorne*, the first Caledonian steamer to be fitted with a triple engine, was not too successful in the important matter of speed and although other factors probably contributed to this disappointment nonetheless it was evident that Captain Williamson decided to play safe in specifying compound engines when ordering the *Duchess of Rothesay* four years later. Fuel economy may well have swung the balance when the *Duchess of Montrose* was built in 1902. One of a batch of intermediate-sized vessels built about the turn of the century, she was intended for general work rather than cruising and had unusually small paddle wheels; so too did the *Mars*, built by the same yard for the Glasgow & South Western in the same year. This feature allowed lighter moving parts, with a reduction in wear and tear and therefore in maintenance costs. The *Duchess of Fife*, however, reverted to larger wheels, and perhaps after all the innovation was not as successful as had been hoped.

The *Duchess of Fife* was a remarkably successful ship and by exceeding her contract speed handsomely while on trial entirely vindicated the choice of triple expansion machinery. A feature of her engine was an arrangement by which live steam from the boiler could be passed directly into the intermediate pressure receiver, by-passing the high pressure cylinders which, in these circumstances, had high pressure steam admitted to each side of the piston and therefore 'floated', doing no work; the engine therefore worked as a compound, with steam at higher pressure than usual in the intermediate and low pressure cylinders. The effect was to allow the steamer to 'spurt' faster for a short time, but for various reasons this could not be done for long periods. For one thing, the increased stresses would have led to unacceptable wear and tear, but a more practical consideration was that the boiler tended to be 'drained' in such conditions, lowering the pressure and 'winding' the steamer. But the device was useful, as it gave a temporary advantage in speed, very handy when racing a rival for a pier, and steam pressure could quickly be rallied while passengers were disembarked after such bursts.

No other Clyde owners used triple expansion machinery despite the success of the *Duchess of Fife*. Compound diagonals remained supreme and even the conservative North British company took it up in preference to the single diagonal when the *Waverley* was built in 1899. It was an inevitable change, since a vessel of her size would have required an unusually large single engine, and would certainly have suffered on an increased scale from all the defects of that type. Once having made the change, the North British did not look back. Not only did the *Marmion* have a compound in 1906, but when the *Lucy Ashton* suffered her breakdown in 1901, she was given a new compound engine instead of simply having her single engine renewed. The *volte face* of the North British was possibly unexpected but it was consistent with the company's policy of building steamers as cheaply as possible without lowering standards, and the *Waverley* of 1899 simply marked the stage at which compounding *had* to come.

The attitude of the Glasgow & South Western Railway was quite different. Faced with substantial inroads upon its traditional traffic as the result of Caledonian competition, the company had little alternative but to disregard cost in laying down new steamers in 1892. Each vessel had to be capable of bettering the speed of its Caledonian counterpart to be sure of regaining for her owners as much of the lost trade as possible. There was no time to experiment—large twin crank compound engines were used in all the new steamers, and the *Neptune, Mercury* and *Glen Sannox* held decisive margins over the first generation of Caledonian steamers, and offset to a degree the disadvantage of Prince's Pier being three miles up-firth from Gourock. There was no such handicap at Ardrossan and the magnificence of the *Glen Sannox* soon drew back to the South Western a great proportion of the Arran traffic which the *Duchess of Hamilton* had so decisively snatched from the *Scotia* two years earlier. But it was done at the cost of extravagant fuel consumption—the most eloquent testimony to this is the fact that when the *Glen Sannox* was requisitioned by the Admiralty for auxiliary duties after the outbreak of the Great War, she had to be returned as unsuitable since she was unable to stay at sea for long enough periods between rebunkering.

OSCILLATING ENGINES AND STEEPLES

David MacBrayne's Clyde fleet stood at the other extreme from the railway steamers. The latter were subsidised quite openly by the lucrative traffic from other branches of railway operation, financial considerations were less important, at least until after 1900, and managements could afford to experiment to some extent. Not so the West Highland owner who had no such resources to draw upon; serving sparsely populated areas, the MacBrayne ships had to be worked economically and conservatism in matters affecting machinery and boilers was extreme. So it was that the *Iona*, *Columba* and *Grenadier* had the long-outdated oscillating machinery typical of vessels built in the fifties and sixties. That the *Iona* should have been thus equipped was understandable, considering that she had been built in 1864, but the use of diagonal engines in a vessel of her size and importance would have excited no comment even then. By 1878, however, the use of oscillating engines in the *Columba* was an anachronism and even though compounding appeared in the *Grenadier's* unique machinery, the continued employment of outmoded machinery of this pattern as late as 1885 was utterly unexpected. Nevertheless, the oscillating engine had certain advantages, not the least of which was extreme economy of space. The oscillating principle involved cylinders free to oscillate on trunnions through which steam was admitted to them; problems of steam tightness were considerable and vessels driven by these engines almost invariably had low pressure boilers.

The period under review saw the final extinction of an even older form of engine, the 'steeple', which was invented by David Napier in the very early years of Clyde steam navigation, and achieved some popularity for several years in the middle of the nineteenth century. The last new ship to be built with machinery of this type was the *Scotia*, in 1880, but in her case the machinery was of double steeple type and no other Clyde vessel was so fitted. In the steeple engine the cylinder was placed vertically, and the piston drove a crosshead above it from which the connecting rod drove on the crankshaft between the cylinder and crosshead. Like the oscillating engine, the steeple was compact but incapable of development in the same way as the newer diagonal type, and it was discarded for Clyde work quite early although examples lingered on in older ships. The final example in service on the Clyde was that of the *Vulcan* which was withdrawn for breaking up in 1902.

A form of machinery that received limited application was the so-called diagonal oscillating type of engine, used in the first and second *Lord of the Isles* and the *Ivanhoe*. In this arrangement there were two oscillating cylinders placed diagonally fore and aft of the crankshaft, driving on to it by means of a common crank. This too was a compact design but its use was confined to the vessels mentioned.

Screw steamers, other than turbines, were almost invariably propelled by compound vertical engines, the best examples being the *Davaar*, *Kinloch* and *Kintyre*, all of which were single screw ships. Twin screw vessels of this type were unknown on the Clyde.

PADDLE WHEELS

Reference has been made to the small paddle wheels of the *Duchess of Montrose* and the *Mars*, both of which were designed for service at moderate speeds, and an attempt was made to provide light, fast-moving machinery. The Lochgoil & Lochlong Company's *Edinburgh Castle* of 1879 was the antithesis of later steamers in this respect. It is stated that Captain Barr, who commanded her until his death in 1906, was the originator of the plan to build this vessel with immense paddle wheels—they were twenty-two feet in diameter—and by this feature she was readily

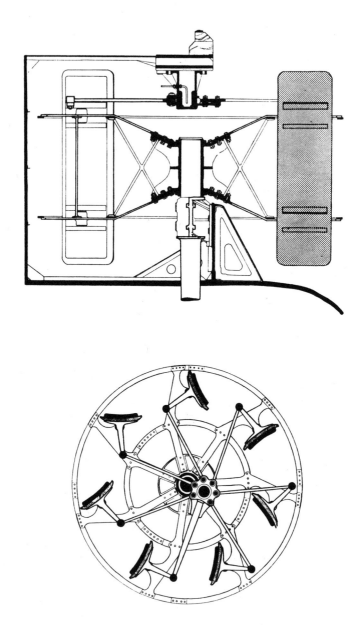

A standard 7-float Caledonian paddle wheel

distinguishable. She was the ugly duckling of the Clyde fleet and unfortunately her huge wheels did nothing to enhance her performance.

Normally, paddle wheels ranged in diameter from sixteen to twenty feet depending on the size of the steamer, the type of machinery fitted, and the services for which they had been designed. The archaic non-feathering type of wheel had utterly vanished by 1889 and all vessels described had the feathering pattern, elm floats being employed. These were cheap and effective and could readily be replaced in the event of damage by floating objects. The most obvious characteristic as far as passengers were concerned was the distinctly audible 'slapping' sound as these floats hit the surface when steamers were sailing. Steel floats replaced wooden ones to some extent later in our period but the older pattern was retained in certain fleets, notably the North British.

A variant of the Navy boiler was the return tube, or 'Scotch', type, in which the hot gases from the furnaces were carried back through the boiler in another row of tubes above the furnaces and thence to the funnel uptake above the firedoors. This was a form of boiler which did not receive wide application on the Clyde, but when the so-called 'double-ended' boiler—in effect two Scotch boilers back to back—appeared, it was taken up by the Caledonian in the *Duchess of Rothesay* in 1895 and by the Glasgow & South Western Railway in the *Jupiter* and *Juno* shortly afterwards. It became virtually the standard boiler for Clyde turbine steamers, with the exception of the *Atalanta*, and resulted in the characteristic twin-funnelled appearance of these vessels although, of course, it could be, and was, designed for one funnel in some ships.

POINTS OF DETAIL

Engine rooms in the turbine steamers presented a complete contrast to those of the paddle steamers. The *King Edward* was unashamedly experimental and little attempt was made to refine machinery details. The controls of paddle vessels were at main deck level, beside the engine room alleyways, but on board the *King Edward* the turbine controls were practically in amongst the engines themselves and passengers looked down on the engineers at work amidst a writhing complex of steampipes from which the steam control wheels projected downwards towards the control platform. The necessary pressure and other gauges were mounted on a panel attached to the pipes. The whole arrangement was crude and reminded one of the work of the late Mr Heath Robinson, but there can be no doubt that the important thing was to see that the turbine principle worked—refinement of detail was secondary. The *King Edward* was never altered in this respect and remained for half a century as a fascinating example of an early turbine engine room.

When the *Duchess of Argyll* came out in 1906 the success of the turbine was beyond question and hers was a much improved layout. The control platform was restored to main deck level, with steam admission valve control wheels placed at waist height for comfortable working, and gauges and counters placed above them on a panel. To those fascinated by the paddlers, the engine room of the *Duchess of Argyll* was a disappointment, but yet it held a strange attraction, and what youngster could fail to be intrigued by the large, polished brass plate bearing the imposing inscription 'Built under Licence from the Parsons Marine Steam Turbine, Coy. Limited, Wallsend-upon-Tyne'? To stand at the engine room hatch and watch the ship cutting effortlessly through the glassy waters of the Firth of Clyde on a fine summer morning, with the steady hiss of the turbines and the noise of auxiliary machinery the only sign of activity, was an unforgettable experience.

Matters of detail design were of great interest in the steamers of the period, particularly in respect of paddle machinery. The compound diagonal type of engine—and for this purpose the three Caledonian triple expansion vessels, and two-cylinder simple expansion steamers come into the same group—quickly evolved into a fairly standard pattern. The earlier examples were assembled on four entablatures of equal size; the *Neptune* and

Mercury of the South Western fleet, and the *Strathmore* and *Kylemore* amongst John Williamson's ships, were so distinguished, but later practice was to dispense with the two inside entablatures in favour of a much larger one. The cranks were placed between the entablatures and the eccentrics of the valve gear outside the small ones.

Older steamers had built-up crankshafts with the straight sides of the crank webs facing inwards towards each other, the tapered or cut-away part facing outwards. Caledonian and Glasgow & South Western steamers built around the turn of the century had solid-forged crankshafts, beautifully made and machined forgings which in themselves were almost works of art. The *Duchess of Montrose, Duchess of Fife,* and *Mars* all had these shafts, amongst others. A high standard of machining and finish was typical of all these vessels; the *Duchess of Rothesay* had 'flats' machined on the sides of her connecting rods so that the fish-belly profile seen in section appeared as parallel sides in plan.

Engine room decoration was usually elaborate. The entablatures gave great scope for complicated lining, while crests and other decorations appeared as centre pieces. Glasgow & South Western entablatures had elaborate floral designs reminiscent of *art nouveau,* and Caledonian vessels sported lions rampant.

This review of machinery practice may conclude by summarising the Clyde steamer of the pre-1914 years as a coal-burning paddle steamer, driven by thermally inefficient machinery. The turbines were in a class by themselves. Looking back, it can be seen as a period in which sea transport, in common with the railways, was dependent entirely upon limitless supplies of good steam coal, obtainable easily and cheaply. It was literally 'dirt cheap' and apart from compounding there was no attempt to refine the conventional reciprocating paddle engine save in matters of detail. We are witnessing a similar process in our own day and age, but the fuel is oil and the engines are diesels— eventually something superior will supersede them. It is impossible to recall the age of the paddle steamer without nostalgia, however. Wasteful both in space and operation, the paddle engine nevertheless remained one of the most impressive creations of the age of steam transport and in its many Clyde variations it exercised a fascination second to none in the field of mechanical engineering.

APPENDICES

Appendix One

COLOURS AND UNIFORMS

The quarter of a century before the Great War was undoubtedly the most colourful in the entire history of Clyde steamer services. Not only were the various fleets at their zenith, but the individual ships were also more attractively painted than at any time before or after. The increasing use of full deck saloons, almost invariably painted in bright pastel shades, or plain white, gave the ships of the period a brighter appearance than their early and mid-Victorian sisters, whose colour schemes were more conservative. A corresponding trend became apparent in steam locomotive liveries in approximately the same period, brighter colours becoming more popular from the turn of the century.

A feature which attracted much interest in the Clyde paddle steamers was the design of paddlebox. Unlike a later generation which turned out vessels disguised to resemble turbine steamers, suppressing paddleboxes as a distinctive feature in favour of plain openings, the designers of the nineties and the Edwardian era produced a whole series of beautiful designs. The finest of all, aesthetically, were those of the Caledonian's later ships, in which a delicate tracery of lattice-work followed the sweeping contour of the paddlebox, forming an attractive background for much decoration. Glasgow & South Western paddleboxes were of very similar type, but in place of the coats of arms which distinguished the Caledonian vessels, they had a circular panel bearing the railway's heraldic device. Blue nameboards and elaborate gilt ornamentation were characteristic of the ships in both fleets.

Both Caledonian and Glasgow & South Western companies favoured white paddleboxes, but other owners preferred black. Amongst these was the North British, which eventually standardised a paddlebox with eight radial vents. David MacBrayne used a paddlebox with five or seven radial vents, applying the design to bought-in steamers as well as new ones. Some steamers remained unique even in their own fleets. The North British *Guy Mannering*, for example, retained her original Wemyss Bay paddleboxes through-

251

out her career at Craigendoran, and into Buchanan ownership. The Caledonian *Duchess of Hamilton* and *Marchioness of Lorne* shared a pattern which never appeared on other Clyde vessels, but was used by Denny on paddlers built for the Thames. The well known *Belle* steamers carried this type, consisting of plain horizontal slots; the outer edge of the box had circular ports, but no name board. The use of this standard Denny paddlebox by Russell & Co, the builders of the *Marchioness of Lorne*, was almost certainly intentional, and probably meant to make the smaller vessel resemble the larger.

Radial vents, however, were more generally popular in the mid-Victorian years; for their principal purpose, that of rapid dispersal of water building up inside the boxes in rough weather or when listing due to crowding on board, they were ideal. Some of the neatest patterns appeared in the privately owned steamers, such as the Campbell vessels sailing to Kilmun. The *Madge Wildfire* kept her original paddleboxes of this type well into Caledonian ownership, but latterly they were replaced by much plainer ones with straight slots. The *Caledonia* and *Marchioness of Breadalbane* suffered a like fate in due course.

In the early nineties the matter of uniform for steamer crews was one for broad interpretation. The *Ivanhoe* had set the pace during the eighties, her owners placing great importance upon a smart and nautical appearance. The deckhands accordingly wore white naval jerseys with the name *Ivanhoe* across the chest, and Captain Williamson and his officers wore naval-type uniform caps and frock coats. The MacBrayne vessels *Iona* and *Columba* were also noted for smartness; being very much in the limelight, the 'Royal Route' tradition demanded something better than the casual turn-out characteristic of the middle years of Clyde steamer history and consequently the captains wore heavily braided caps of gorgeous appearance, while officers and crew were also well turned out. This tradition also extended to the *Lords of the Isles*, which were both in the same tourist class as the MacBrayne steamers.

Amongst the older skippers, however, former traditions died hard, and the old, high bowler hat was much favoured, Captain Barr of the *Edinburgh Castle* being thus noted. Wearing apparel of the crews of older ships was not particularly nautical, and could be described as practical but nondescript. The modern style of uniform really came into its own with the railway fleets of the nineties. The Caledonian led the way, followed by the South Western, while the North British owners tended to be more conservative in this, as in other matters. By 1914, however, the change was complete, and smart uniforms for officers and crew alike were universal.

DETAILED LIVERY NOTES

Caledonian Steam Packet Co Ltd

The colour scheme chosen for this fleet was redolent of naval vessels and steam yachts. Based on the colours used for the *Ivanhoe*, there were however several modifications.

Funnels were creamy yellow, without black tops. Hulls were painted a very dark shade of blue. Above a green underbody, there was a broad white boot-topping. In later years the green was changed to the more conventional dark red. Two gilt lines were carried round the hull and sponsons just below mainrail level, from bow to stern, with extra gilt ornamentation on the stern in some cases. Names and port of registry were likewise in gilt letters.

Saloons were painted a delicate pale pink shade, as were the sponson housings, but panelling between saloon windows was pale blue. Wash boards at promenade deck level were painted white.

The paddleboxes of Caledonian steamers were their special glory. Those ships named after royalty and nobility were given large, hand-carved armorial bearings for crests, beautifully painted in full heraldic colours. Against the white background of the boxes, they made a brave show and were much admired. Those of the *Caledonia*, *Duchess of Hamilton* and *Marchioness of Lorne* were preserved, and the first and last-named vessels' crests may still be seen on display at Wemyss Bay and Gourock. All steamers save the *Duchess of Hamilton* and *Marchioness of Lorne* had blue nameboards with gilt lettering, and paddleboxes generally were heavily ornamented in gilt paint.

Deckhouses were generally varnished teak, but the *Hamilton* and the *Lorne* were unique in their white deckhouses, panelled in each case in pale blue and pale pink. Lifeboats were always white, with white covers.

Variations occurred in the foregoing general scheme, including specifically

(1) Omission of gilt lines from the hull of the *Duchess of Argyll*;

(2) Brown, or varnished, wash boards on the *Duchess of Fife* in her early years.

North British Steam Packet Co and North British Railway

Change of ownership in 1902 did not affect the colour scheme of the steamers sailing from Craigendoran. Beside the brighter, fresher colours of the other railway fleets, the North British vessels were distinctly Victorian in appearance, but in course of time their distinctive livery was eventually to outlast those of its rivals by many years.

The Craigendoran funnel was red, with a broad white band, and black top. The narrow stay ring between red and white was always painted black.

The hull was black, with dark red underbody, and a thin white line separated red from black. In earlier years, the white band had been much broader, similar to Caledonian practice. Two gilt lines ran from bow to stern, and round the sponsons, and there was a good deal of gilt-work at the stern, including the vessel's name and port of registry.

Saloons and sponson houses were cream, panelled in light brown, and wash boards above them were white, giving a pleasant lightening effect.

Paddleboxes were black. Latterly, all steamers had the standard Inglis form of box, with eight radial vents, picked out in gilt, but earlier ships had more vents (eg the *Lady Rowena*, with ten) and there was a vertical ornamental pillar, heavily gilded, above the crest, dividing the vents into two equal groups. The crest was in the form of a hand-carved bust of the Sir Walter Scott hero or heroine after which the vessel was named, supported by a garland of ornamental thistles.

Deckhouses were varnished teak in all cases, and lifeboats white.

Glasgow & South Western Railway

After a very brief experiment with a black hull, combined with a red and black funnel, this company adopted, and retained throughout the period, a very beautiful colour scheme inspired, it is understood, by that of the Union Castle liners.

The funnel was scarlet, with black top.

The hull was light grey, with dark red underbody, and plain white saloons and sponson housings. Dark brown lines were painted round the hull at mainrail level, at the promenade deck, and at lower sponson level, from stem to stern.

Paddleboxes were white, with much ornamentation, including the South Western company's crest painted on a circular panel on all new ships. Ships purchased secondhand, except the *Juno*, retained their original paddleboxes.

Deckhouses were varnished teak, and lifeboats were white, with white covers.

David MacBrayne and David MacBrayne Ltd

The vessels of the MacBrayne fleet had black hulls, with red underbody, the two colours being separated by a thin white line. Gilt lining was carried round the hull from stem to stern below mainrail level, and in the case of all three vessels described in this book there was unusually elaborate gilt scrollwork at bow and stern. The *Grenadier* had a figure-head, sharing this unusual feature with the three screw vessels of the Campbeltown company, and the *Culzean Castle*.

MacBrayne's standard paddlebox had five or seven radial vents and the *Columba* was particularly noted for her scrollwork and the ornamental thistle twining from the letter 'U' in the nameboard. All vessels had black paddleboxes.

Saloons were officially 'stone colour', which appeared as a creamy shade, but in the early years of the Great War the colour was a light blue, which may have been introduced before 1914, although precise evidence is lacking.

Deckhouses were varnished teak, and lifeboats white.

The MacBrayne funnel was bright red, with a black top, and thin black hoops. For an interesting discussion of the origin of these colours, and the early relationship between the Cunard Company and MacBrayne, readers are referred to Duckworth & Langmuir's *West Highland Steamers*. The black hoops were painted red from about the time of the formation of the limited company, so that the colours became simply red, with black top.

Glasgow & Inveraray Steamboat Co Ltd and
Lochgoil & Lochlong Steamboat Co Ltd

The two *Lords of the Isles* shared a beautiful colour scheme with the less imposing ships of the Lochgoil route. The funnel was red, with two narrow white bands enclosing a black one of equal width, and a black top. The second *Lord of the Isles* was the last Clyde steamer to retain the old-fashioned, brightly polished copper waste steam pipes which, on most vessels, were painted in the funnel colours.

The hull was black, with red underbody and a thin white line dividing red from black at waterline level. The usual gilt lines were carried round the hull and sponsons.

Saloons and sponson housings are thought to have been white, at least in later years, but the fashion in earlier years was to use a pastel shade rather than white, and the possibility that the Inveraray and Lochgoil vessels were so painted cannot be ruled out. Specific evidence on this point would be welcomed by the author.

Paddleboxes were black, with radial vents, and rich gilding.

Lifeboats were white, and deckhouses varnished teak.

There was no change of colours when the *Edinburgh Castle* and *Lord of the Isles* passed into the ownership of Turbine Steamers Ltd in 1912.

Alex. Williamson's 'Turkish Fleet'; and Captain Buchanan, and Buchanan Steamers Ltd

The older Clyde vessels without full deck saloons tended to look dull in comparison with their younger sisters, chiefly due to the lack of white paint. This was particularly true of the older Williamson and Buchanan steamers, which shared a livery of black hull, relieved by gilt lining, and a black funnel, with a broad white band. The short deck saloons of the *Sultana* and *Marquis of Bute* were not large enough to lighten their appearance noticeably. The *Viceroy*, after rebuilding and just before purchase by the Glasgow & South Western Railway, had light deck saloons, as did Buchanan's *Scotia*. It is not certain that white was used, and old photographs reveal quite plainly that in the case of the *Isle of Arran* another colour was employed, contrasting perfectly plainly with the white wash boards.

Paddleboxes of pre-1891 Williamson and all Buchanan vessels were black, with the usual ornamentation, the *Scotia* being especially noteworthy in this respect. Just before the outbreak of war, Buchanan steamers sailed for a season or two with white paddleboxes, and it seems that by this time the saloons were also white.

Captain John Williamson

Captain Williamson inherited the colours of his father's 'Turkish Fleet', but modified them in course of time. The paddleboxes became white instead of black in 1896, and the black funnel with broad white band gave way to an all-white funnel, with black top, in June 1898.

The hull was black, with saloons latterly white, but some other colour was used earlier. Gilt lining on the hull was later abandoned.

Deckhouses were varnished teak, and lifeboats white, but the *Queen Empress* had varnished boats.

Turbine Steamers Ltd

The Turbine Syndicate and Turbine Steamers Ltd inherited the basic Williamson colours, modified in certain respects as befitted newer ships.

The funnels were white, with black tops.

Hulls, black, with red underbody and no lining. Saloons and wash boards were white. The painting specification called for brown lettering for the name *King Edward* on the bow plating of the pioneer turbine vessel.

Deckhouses were varnished teak.

Although the *King Edward*'s lifeboats were white, with white covers, those of the first *Queen Alexandra* were plain-varnished wood. So also were those of the second *Queen Alexandra* in her early years.

Campbeltown & Glasgow Steam Packet Joint Stock Co Ltd

These screw vessels had black funnels, with a broad red band, practically half the depth of the funnel. Hulls were black, with pink underbody, and a thin white dividing line between black and red at the waterline.

Saloons and certain parts of the superstructure were white, but deckhouses were the usual varnished teak. There was a good deal of gilt ornamentation at bow and stern, but no lining round the hull. Lifeboats were white.

Wemyss Bay Co (Captain Alexander Campbell)

These steamers were painted in a rather stark livery consisting of black hull, white saloons *(Victoria)* and white funnel with black top. The older ships' short saloons were varnished teak. Paddleboxes were black, with gilt ornamental work, and the hulls had gilt lining from bow to stern.

Frith of Clyde Steam Packet Co Ltd

The *Ivanhoe* was the first modern Clyde steamer to have yellow funnels; although these appeared all-yellow, photographs reveal that the top rings were black. The hull and paddleboxes were black, with broad white boot-topping and the usual gilt lining and ornamentation. Saloons were a light colour, but no record remains of whether they were white, or a pastel shade.

Miscellaneous

Dawson Reid turned out the *Duchess of York* in a lurid and uncharacteristic colour scheme of grey hull, with white saloons, grey paddleboxes, and yellow funnel, with two narrow white bands enclosing a red one. It was probably the most tasteless Clyde livery of the whole period, and did little to improve the old *Jeanie Deans.*

Captain A. Cameron ran the *Madge Wildfire* in 1912 with a green hull, and red funnel with black top, a not unpleasing combination.

The yellow, or buff, funnel with black top, so familiar in later years, made only brief appearances on the Clyde in the period— it was applied to the *Victoria* in the mid-nineties, to the *Culzean Castle* for a period, and to the *Madge Wildfire* in 1911.

The *Ivanhoe* sailed at first with all-white funnels in the ownership of the Firth of Clyde Steam Packet Co Ltd, but later there were narrow black tops, of about half the usual depth.

Appendix Two

THE CLYDE STEAMER FLEET IN 1914

Buchanan Steamers Ltd

P.S. *Isle of Cumbrae*	1884
P.S. *Isle of Skye*	1886
P.S. *Isle of Arran*	1892
P.S. *Eagle III*	1910

Caledonian Steam Packet Co Ltd

P.S. *Caledonia*	1889
P.S. *Marchioness of Breadalbane*	1890
P.S. *Duchess of Hamilton*	1890
P.S. *Marchioness of Lorne*	1891
P.S. *Duchess of Rothesay*	1895
P.S. *Duchess of Montrose*	1902
P.S. *Duchess of Fife*	1903
Tr.S.S. *Duchess of Argyll*	1906

Captain A. Cameron

P.S. *Lady Rowena*	1891

Campbeltown & Glasgow Steam Packet Joint Stock Co Ltd

S.S. *Kinloch*	1878
S.S. *Davaar*	1885

Glasgow & South Western Railway

P.S. *Neptune*	1892
P.S. *Mercury*	1892
P.S. *Glen Sannox*	1892
P.S. *Minerva*	1893
P.S. *Glen Rosa*	1893
P.S. *Jupiter*	1896
P.S. *Juno*	1898
P.S. *Mars*	1902
Tr.S.S. *Atalanta*	1906

David MacBrayne Ltd

P.S. *Iona*	1864

P.S. *Columba*	1878	**Turbine Steamers Ltd**	
P.S. *Grenadier*	1885	Tr.S.S. *King Edward*	1901
		Tr.S.S. *Queen Alexandra*	1912
North British Railway		P.S. *Ivanhoe*	1880
P.S. *Lucy Ashton*	1888	P.S. *Lord of the Isles*	1891
P.S. *Dandie Dinmont*	1895		
P.S. *Talisman*	1896	**Captain John Williamson**	
P.S. *Kenilworth*	1898	P.S. *Benmore*	1876
P.S. *Waverley*	1899	P.S. *Kylemore*	1897
P.S. *Marmion*	1906	P.S. *Queen Empress*	1912

Appendix Three

CLYDE PIERS AND FERRIES 1889 to 1914

GLASGOW AND UP RIVER
Broomielaw
Bridge Wharf
Partick Wharf
Govan
Renfrew
Erskine Ferry
Bowling
Dumbarton
Dumbarton Old Quay
Port Glasgow
Custom House Quay
(Greenock)

RENFREWSHIRE
Prince's Pier (Greenock)
Gourock
Wemyss Bay

AYRSHIRE
Largs
Fairlie
Portincross
Montgomerie Pier (Ardrossan)
Winton Pier (Ardrossan)
Troon

Ayr
Girvan

WIGTOWNSHIRE
Stranraer

DUNBARTONSHIRE

The Gareloch
Row
Shandon
Rahane (ferry)
Rosneath
Clynder
Barremman
Mambeg
Garelochhead

The Firth
Craigendoran
Helensburgh
Kilcreggan

Loch Long
Cove
Coulport
Portincaple (ferry)
Arrochar

COUNTY OF BUTE

Isle of Arran
Brodick
Lamlash
King's Cross (ferry)
Whiting Bay
Whiting Bay (ferry)
Kildonan (ferry)
Blackwaterfoot (ferry)
Machrie Bay (ferry)
Pirnmill (ferry)
Lochranza
Corrie (ferry)

Isle of Bute
Rothesay
Craigmore
Kilchattan Bay
Port Bannatyne

Isle of Cumbrae
Keppel
Millport

ARGYLLSHIRE

Loch Long and Loch Goil
Blairmore
Ardentinny (ferry)
Carrick Castle
Douglas
Lochgoilhead

The Holy Loch
Strone
Kilmun
Ardenadam
Hunter's Quay

Cowal
Kirn
Dunoon
Innellan
Toward

Kyles of Bute
Colintraive
Ormidale
Tighnabruaich
Auchenlochan
Kames
Ardlamont (ferry)

Loch Fyne
Tarbert
Ardrishaig
Otter Ferry (ferry)
Strachur
Crarae
Furnace
Inveraray

Kintyre
Skipness
Carradale
Campbeltown

Note: The inclusion of a pier or ferry in the above list does not signify that it was in constant use throughout the period, but in each case cited there is evidence to prove that it was used at some point. Some piers were closed quite early (eg Clynder) and others opened late (eg Portincross), while in the case of Whiting Bay a pier replaced the existing ferry in 1899. In other cases the services were sporadic, as in the cases of Port Glasgow and Dumbarton Old Quay.

Appendix Four

P.S. NEPTUNE CRUISES JULY 1896

July 2   Ayr to Troon, Ardrossan, Millport, Largs, Dunoon, Lochgoilhead.
         Evening Cruise—Ayr to Lamlash.
     3   Ayr to Troon, Ardrossan, Largs and Arrochar.
     4   Ayr to Troon, Ardrossan, Largs and Hunter's Quay to follow Corinthian Yacht Club Regatta.
     5   SUNDAY—NO CRUISE
     6   Greenock to Dunoon, Innellan, Rothesay, Craigmore, Largs, Millport and Ayr.
     7   Girvan to Ayr, Troon and Rothesay.
         Evening Cruise—Ayr to Girvan.
     8   Ayr to Troon, Ardrossan, Lamlash, Brodick, Tighnabruaich and Ormidale.
     9   Ayr to Troon, Ardrossan, Largs and Arrochar.
         Evening Cruise—Ayr to Pladda.
   10   Ayr to Troon, Ardrossan, Millport, Largs, Dunoon and Lochgoilhead.
   11   Ayr to Troon, Ardrossan, Largs and Rothesay to follow Royal Northern Yacht Club Regatta.
         Evening Cruise—Ayr to Culzean Bay and Turnberry.
   12   SUNDAY—NO CRUISE
   13   Greenock to Dunoon, Innellan, Rothesay, Craigmore, Largs, Millport and Ayr.
   14   Rothesay to Largs, Millport, Ardrossan, Troon, Ayr and Stranraer.
   15   Ayr to Troon, Ardrossan, Millport, Largs, Rothesay, Tighnabruaich, Auchenlochan and Kames.
   16   NO CRUISE
         Evening Cruise—Ayr to Brodick.
   17   NO CRUISE
   18   Ayr to Troon and Brodick (forenoon); Troon to Ayr and Girvan (afternoon)
   19   SUNDAY—NO CRUISE
   20   Greenock to Dunoon, Innellan, Rothesay, Craigmore, Largs, Millport and Ayr.
   21   Ayr to Troon, Ardrossan, Largs and Arrochar.

22 Ayr to Troon, Ardrossan, Millport, Largs, Dunoon and Lochgoilhead.
   Evening Cruise—Ayr to Lamlash.
23 Ayr to Troon, Ardrossan, Millport, Largs, Rothesay, Tighnabruaich, Auchenlochan and Kames.
24 Ayr to Troon, Ardrossan, Lamlash, Brodick, Tighnabruaich and Ormidale.
25 Girvan to Ayr, and Troon, and Round Arran.
   Grand Evening Cruise from Troon and Ayr to Brodick.
26 SUNDAY—NO CRUISE
27 Greenock to Dunoon, Innellan, Rothesay, Craigmore, Largs, Millport and Ayr.
28 Largs to Millport, Ardrossan, Troon, Ayr and Stranraer.
   Evening Cruise—Ayr to Largs.
29 Ayr to Troon, Ardrossan, Largs and Arrochar.
30 Ayr to Troon, Ardrossan, Millport, Largs, Dunoon and Garelochhead.
31 Ayr to Round Arran Cruise, returning via Corrie, Brodick and Ardrossan.

Appendix Five

THE TURBINE SYNDICATE AGREEMENT, 1901

AGREEMENT FOR THE BUILDING OF A SCREW STEAMER FOR COAST AND CHANNEL SERVICE

The parties to the Syndicate are :
Messrs William Denny & Brothers, Leven Ship Yard, Dumbarton.
Messrs The Parsons Marine Steam Turbine Company, Limited, Turbinia Works, Wallsend-on-Tyne.
Captain John Williamson, 7 Bridge Wharf, Glasgow.

———

1. Each of the above parties in this Syndicate are interested in the proportion of one third in the undertaking, including the construction and running of the vessel for the next summer season, or for such longer period as may be arranged until the vessel is sold.

2. Messrs William Denny & Brothers, and the Parsons Marine Steam Turbine Co Ltd, are to be the joint constructors of the vessel and machinery, the work being divided as follows :

Messrs William Denny & Brothers build and equip the hull, supply and fit the boiler, pipes, fans, flooring plates, ladders, handrails, &c, &c, in fact all the work with the exception of that enumerated in Clause 3. The work to be of the best and suitable for such a service as that between ports on the Clyde and Campbeltown, as well & as far as possible, suitable for other similar services with a view of advantageous sale to other parties, if desired.

3. The Parsons Marine Steam Turbine Co Ltd, construct and fit on board at Dumbarton (rough labour excepted) the following propelling machinery :

1. The main steam turbines.
   (a) One high pressure on centre shaft.
   (b) Two low pressure and reversing turbines on the outer shafts.
   (c) Two sets of two compound air pumps each worked off low pressure turbines.
2. Two surface condensers of 3,000 sq ft cooling surface each.
3. One feed heater.
4. Two sets of circulating pumps, & auxiliary air pumps.
5. All piping, valves, connected with the above machinery, but not the sea fittings on ship's side or sea cocks.
6. Shafting, propellers, shaft liners, stern tubes, but not the brackets for carrying the shafts.

4. The cost of Messrs William Denny & Brothers' part of the work, in completing the vessel in every respect ready for commission, including Steward's furnishings for the vessel, to be £24,200, (say twenty-four thousand two hundred pounds sterling).

5. The cost of the Parsons Marine Steam Turbine Coy's part of the work to be £8,000 (say eight thousand pounds sterling).

6. The Syndicate to provide equally among them £800 (say eight hundred pounds sterling) floating capital to be applied to meet the expenses for docking, painting after completion, trial trips, legal and preliminary expenses. Should any balance be left out of this £800, after the vessel is on her service, it is to become the property of the parties to this Syndicate in proportion to their respective interests.

7. The cost of the vessel to the Syndicate is thus not to exceed £33,000 (say thirty-three thousand pounds sterling).

8. No extras to be charged by the Contractors to the Syndicate.

9. The following are appointed a Committee to deal with the finances of the Syndicate :

Captain John Williamson
Archibald Denny, Esq.
C. J. Leyland, Esq.

10. During construction of the vessel, the Committee will pay to Messrs William Denny & Brothers, and the Parsons Marine Steam Turbine Co Ltd, the instalments on their respective contracts as they became due, as per schedule of scheme of instalments.

11. Captain Williamson to trade the vessel on behalf of the Syndicate until sold as under clause 1, on the Fairlie and Campbeltown run.
Profits or losses to belong to, or be borne by, the parties to this Syndicate in proportion to their holdings.

12. In the event of Captain Williamson's death, or his being unable to carry on the management, the majority in value of the Syndicate to have power to appoint a Manager, but the steamer to continue to run on the route stated in clause 11.

13. Captain Williamson as Managing Owner undertakes to run the vessel during the first season without remuneration.

14. After the first season's running, Captain Williamson to have the option for two months after the 31st October of purchasing the vessel at the cost to the Syndicate, viz : £33,000 (say thirty-three thousand pounds sterling).

15. Failing the exercise of this option, the vessel to be sold to the best advantage, each party receiving the profit, or bearing the loss in proportion to their holding in the Syndicate.

16. If any floating capital other than the £800 previously provided for is required in the first instance to start the vessel in her trade, the same to be provided equally by the parties to this agreement.

17. Arbitration clause as usual.

18. The schedule of scheme instalments to be paid under clause 10, to be as follows :
William Denny & Brothers, Ship No "651".
1st instalment when vessel is framed, and beamed, bulkheads in place, and all internal work rivetted.
2nd instalment when vessel is launched.
3rd instalment when the vessel is delivered after trial.
Messrs The Parsons Marine Steam Turbine Co Ltd, Engine No. 8.

1st, 2nd, and 3rd instalments at the same time as those of Messrs William Denny & Brothers.

(*Sgd*) Wm. Denny & Brothers   *8th April, 1901*
  John Williamson   *10th April, 1901*
   pro THE PARSONS MARINE STEAM TURBINE CO (LIMITED)
  C. J. Leyland *6 May, 1901*
  Charles A. Parsons *6 May, 1901*
   Directors.

Appendix Six

CLYDE COAST HOLIDAY ACCOMMODATION

Glasgow Herald Advertisements, June 1898

CLYNDER—2 Public, 6 Bed Rooms, bath, piano, boat; shore and pier; July £15.—Turner.

CLYNDER—7 Apartments, 7 beds; near shore and pier; £10 monthly.—Turner.

KILGREGGAN—Good House on shore; 6 apartments; good view; June £5.—Kerr, agent.

ARRAN (Southend)—Farm House, 3 Apartments, 4 beds; highly recommended, July, August; £5 monthly; luggage free.—Rae, 72 Paisley Road West.

ARRAN (Blackwaterfoot)—Double Bedded Room, with cooking stove; £1 weekly.—Drysdale, 83 Renfield St.

BUTE—The Hermitage, Ascog; charming bungalow with beautiful grounds; £20 monthly.—Apply Estate Office, Rothesay.

LAMLASH—Room, Kitchen, 2 beds, now till 28th, rent, £1 15s; also 3 Rooms, attendance.—Montgomery.

LAMLASH—House, 3 Bed rooms, parlour, kitchen (6 beds), for June; rent £3.—Post Office.

MILLPORT—2 Rooms and Kitchen; splendid view; water in house; July, £7.—Mrs. Stevenson, 657 Dumbarton Road, Partick West.

ROTHESAY (Glenhead, Ballochgoy)—Attic Flat. 2 Up, 5 apartments; splendid view; August £6—Forson, 13 Newhall Terrace, or 35 Newhall Street, Glasgow.

DUNOON—3 Apartments (4 beds); own gate; June, 15s weekly.—McGregor, stationer, Harmony Row, Govan.

HUNTER'S QUAY—Dunmore, comfortably furnished; 7 apartments (5 beds), hot bath, gas, organ.—Apply there.

Appendix Seven

SPECIFICATION OF PADDLE STEAMER
LADY ROWENA

Built for the North British Steam Packet Company, 1891.

Specification of Steamer : 200 feet × 21 feet × 7′ 3″ feet [*sic*]

Building under Contract with
Messrs. Hutson & Corbett, Glasgow.

General Description.

Length between perpendiculars	200 feet
Breadth	21 ,,
Depth moulded	7.3 ,,
Rise of floors amidships	4 inches
Round of beam	8 inches
Height of bulwarks at lowest	3′ 2″
Draught not to exceed	4 feet

Built to Board of Trade requirements for a number 3 summer certificate, exclusive of new life saving apparatus regulations.

To have a sitting saloon on main deck aft, full breadth of vessel and 40 feet long.

A dining saloon on main deck forward with passage between it and the bulwarks, 35 feet long.

Refreshment room below main deck forward having in it rooms for Engineer and Purser also Steward's pantry.

Crew accommodation below main deck forward of Refreshment room.

A promenade deck over Saloons about 135 feet long.

After sponsons to be long and broad, to carry ladies' cabin with two water closets on port wing and gentlemen's cabin with water closets and urinals on starboard wing.

Forward sponson to carry water closets and urinals on starboard wing and cooking galley and Steward's store on port wing.

Captain's deck house of teak wood to be placed on promenade deck in front of chimney [*sic*!] with bridge leading from it to paddle boxes.

Steering gear with binnacle and Chadburn's telegraph to be placed on top of Captain's deck house.

Details

Rudder . . . The Rudder to communicate by chains and rods with Steering Gear on Captain's house. A spare tiller to be fitted.

Frames. To be 24 inches apart in Engine and Boiler space, doubled in bulkheads and for two frames at paddle shaft also at paddle beam bracket. To be double for two feet below under turn of bilge. In centre the frames to be $2\frac{1}{2}''$ x $2\frac{1}{2}''$ x 5/16'' and at fore end and aft end to be $2\frac{1}{4}''$ x $2\frac{1}{4}''$ x $\frac{1}{4}''$ and spaced thirty inches apart. The frames alongside of after saloon to be carried up to saloon deck beams.

Engine and Boiler Seats to be as directed and of at least the same strength as those on the Company's Steamer 'Jeanie Deans'.

Plating. Garboard strake to be 5/16'' thick for 100 feet amidship and $\frac{1}{4}''$ at ends. Bottom and side strakes to be $\frac{1}{4}''$ for 100 feet amidship and 3/16'' at ends. Shear strake to be 3/8'' for a length of 12 feet forward and abaft of Engineroom bulkheads and $\frac{1}{4}''$ for remainder of length. . . .

Hull Rivetting. Keel, Stem, Stern post and all butts to be double rivetted. Shell plating for 110 feet amidship to be double rivetted, the remainder to be single rivetted.

After Bulwarks. To be of steel 3/16'' thick with $2\frac{1}{2}''$ x $2\frac{1}{2}''$ x $\frac{1}{4}''$ angle on top for rail. Forward Bulwarks to be of steel plate also with alternate frames carried up to rail.

Deck Stanchions as required, all of tube $2\frac{1}{2}''$ diam. with solid ends welded on. Those in after saloon and dining saloon to be covered with carved wood.

Rail Stanchions, Rods etc. Those around saloons and bridge deck, also on quarter deck to be of suitable size with good soles, secured through covering board and paddle boxes by two screw bolts. The stanchions to be 4 feet apart with 2 rods $\frac{5}{8}''$ running through them and *all* covered with galvanized wire netting. The stanchions and rods also to be galvanized.

SPECIFICATION
OF
DIAGONAL SURFACE CONDENSING ENGINE AND BOILER
FOR PADDLE STEAMER
200' x 21' x 7' 3'' for the *North British Steam Packet Company*

Cylinder. To be designed 50 inches diameter with a stroke of 72 inches. To be cast of good hard close-grained metal free from all defects, and smoothly bored and planed. To have spring loaded escape valves on top and bottom of cylinder for taking off the water, all necessary drain cocks and pipes, and to be fitted with indicator cocks and pipes. To have two slide valves, one on each end of cylinder, with ports at each end to save steam, and wrought by eccentric, connected to steam starting gear.

Cylinder Cover. To be of good hard cast iron, strengthened by stiffening ribs, and polished on flanges.

Piston. To be of the box pattern strongly ribbed. Junk and packing rings to be truly turned and scraped to piston.

Valve Gear. Eccentric pulley to be cast singly with flanges on outside, and strap to have large bearing surface. Eccentrics to be of malleable iron, polished bright, with T end bolted to strap at top, and fitted with stoup and cutter at bottom end. Valve spindle to be carried down through valves with loose collars at top, and double nuts at bottom end. Valve spindle to have guide above gland.

Starting Gear. To have a suitable steam cylinder for starting or stopping engines, and all handles to be carried up to a neat column with brass index plate on top, fixed on starting platform.

Throttle Valve. To be close to cylinder casing. Chest to be lined with brass and have brass seat and spindle. Spindle to be carried up to starting platform.

Condenser. To be surface condensing cast of good strong cast iron, well ribbed, and fitted with brass tube plates, and solid drawn brass tubes of best make ¾ inch external diameter, and packed at each end by cotton rings and secured by screwed brass glands capped to prevent tubes going on end. To have about 1700 square feet of cooling surface. To be fitted with chest and pipes for waste steam from boiler.

Air Pump. To be wrought from main engine by bell crank, levers and rods, with necessary bolts and brasses. . . .

Framing. To be of strong cast iron, securely bolted to cylinder and sole-plate with turned bolts and nuts. Guides to be cast on. To have two large bearings with best gunmetal bushes, malleable iron covers, strong bolts and lubricators.

Guide Blocks. To have cast iron guide shoes with large bearing surfaces.

Soleplate. To be of strong cast iron bolted to condenser and framing, strongly put together, and well secured to stools of ship.

Shafts, Cranks, etc. Shafts, cranks, piston and connecting rods all to be of best forged iron, and have brass bushes where required. Bolts, spindles and rods in contact with sea water to be of brass or malleable iron covered with brass.

Paddle Wheels. To be on the feathering principle about 19′ 0″ in diameter, with eight floats of elm on each wheel about 8 feet 6 inches by 3 feet, and centres 4 feet 6 inches in diameter of cast iron. The arms and rings to be of malleable iron strongly made and fitted together. The bushes to be of brass and pins covered with brass, and the arms, rings, and bolt holes in centre to be rimmelled out, and fitted with well fitted turned bolts.

Telegraph. Three Chadburn's telegraphs for engineroom communication, one from top of Captain's house and one from top of each paddle box.

Boiler. To have one haystack multitubular boiler 14 feet 6 inches diameter by 15 feet high with four furnaces with paped mouths and malleable iron doors and cast iron dampers. The shells and all other plates in the boiler to be of mild steel of Dalzell or Newton brands, and the tubes, which are to be $2\frac{1}{4}$ or $2\frac{1}{2}$ inches diameter by No. 9 W.G., and stays to be of malleable iron. All holes to be made fair by rimmelling before being rivetted, no drifts to be used, and the boiler to be thoroughly caulked inside and out. To be stayed and rivetted to Board of Trade requirements for a working pressure of 55 lbs per square inch, and to be proved by hydraulic pressure before leaving works to the satisfaction of the Board of Trade surveyor to 110 lbs per square inch. To be fitted with all necessary man and sludge hole doors, furnace bars and bearers.

Funnel. Outside funnel to be about 5 feet 3 inches diameter by 23 feet high and inside one about 4 feet in diameter, to be flush rivetted. To be fitted with damper wrought from engine room, and manhole doors at bottom.

Boiler Mounting. Boiler to be fitted with all necessary mounting including the following:

Spring loaded safety valve with copper waste steam pipe led up funnel, and easing gear wrought from engine room.

Stop valve with main steam pipe to engine, and copper internal pipe carried well up inside boiler to get dry steam.

Blow off cocks on boiler and ship's side with copper pipes between.

Organ whistle with stop valve and copper pipe led up funnel.

Stokehold. To be covered with malleable iron chequered plates, close down to floor of ship. Ashpans to be covered with firebrick scones 1″ thick.

Ventilators. Four galvanized iron ventilators abt. 24 inches diameter for stokehold.

Appendix Eight

GLASGOW & SOUTH WESTERN RAILWAY

MEMORANDUM

AS TO

PASSENGER TRAFFIC

GLASGOW AND CAMPBELTOWN

There is a considerable traffic between Glasgow and Campbeltown, including the west side of Arran, as well as from Stations in Ayrshire to these places—and this traffic is capable of great development if it was fostered and encouraged more than it is at present.

The steamer running from Fairlie to Campbeltown in connection with this Company belongs to Captain John Williamson. It makes a daily run there and back during the Season—which lasts from June till September—the time occupied in the journey from Glasgow to Campbeltown being 3 hours 35 minutes and *vice versa* 4 hours 10 minutes.

The Caledonian Company have a connection via Wemyss Bay to Campbeltown, the water service being performed by the Campbeltown Company's Steamer 'Davaar', the time occupied from Glasgow (Central) to Campbeltown via *Wemyss Bay* and the 'Davaar' being 4 hours 5 minutes, and the return journey 4 hours 23 minutes.

The Caledonian Company have recently expedited their service and are competing more and more every day for this traffic.

Undernoted is a statement of the number of Passengers and Steamboat and Railway proportions on through booked traffic between Glasgow and South Western Stations and Campbeltown, Lochranza, Machrie Bay and Millport *via* Fairlie in connection with Captain John Williamson's Steamer for the Season 1898.

VIA FAIRLIE

	SINGLE			RETURN			Total	Boat Prop'n	Rail Prop'n
	1st Cl.	3rd Cl. & Sal.	3rd Cl. & Stge	1st Cl.	3rd Cl. & Sal.	3rd Cl. & Stge	Pass.	£ s d	£ s d
Campbeltown	102	195	190	808	7,104	6,521	14,920	1,027 – –	719 – –
Lochranza	26	68	57	126	1,198	1,014	2,489	177 – –	125 – –
Machrie Bay	7	47	37	43	989	573	1,696	138 – –	88 – –
Millport	2	59		63	5,001		5,125	65 – –	262 – –
Gross Totals	137	369	284	1,040	14,292	8,108	24,230	1,407 – –	1,194 – –

NOTES—Returns are doubled.

A few tickets issued at North British Railway Stations not yet settled.

It will be seen from this that there is a considerable traffic presently flowing and, as I have said might be further developed.

Since the 'Strathmore' started on the service this year something serious has occurred with her boilers, and we do not know the day when she may break down, which will result in the suspension of our service to and from Campbeltown, greatly to the detriment of the Fairlie route.

We have no Steamer that can be put on the route, as we are debarred from running our boats to Campbeltown, and I do not see how we can charter one of our Steamers for the purpose.

If the service therefore is to be maintained Captain Williamson will require to charter an outside Steamer or to purchase one outright, and the question is how far are we to assist him in so doing? We cannot expect a private Owner to run all the risk of purchasing a boat and putting it on to the Campbeltown run unless he is to be encouraged by the Railway Company, as it is to the Railway Company's own interest that that traffic should be developed. The day is not far distant when we shall be applying for an extension of our present *limited* Steamboat Powers. Campbeltown will be one of the ports to which we shall apply for liberty to run, and therefore in the meantime we should do everything we can to keep up a connection between Fairlie and Campbeltown as against the competing route from Glasgow *via* Wemyss Bay to Campbeltown.

There are three methods suggested under either of which the private owner could be assisted :

(1) To sell one of our Steamers, say the 'Minerva', at the figure at which it stands in the Company's books, viz : £10,390 and to buy her back at the end of the Season at the same figure.

(2) To pay the interest on the cost of a new Steamer to be pur-

chased by Captain John Williamson for a period of—say 3 years.

(3) To subsidise for a period of 3 years a private Steamer running from Fairlie to the extent of—say £500 per annum.

Glasgow,
23rd June, 1899   (Int). D.C.

Appendix Nine

THE CALEDONIAN STEAM PACKET COMPANY, LTD

FINANCIAL ACCOUNTS

FOR THE

YEARS ENDED

31st DECEMBER, 1892 and 1893

REVENUE ACCOUNT

	1893 £	1892 £	Difference Increase £	Difference Decrease £
Receipts.				
Cash collected on board Steamers	9,274	8,678	596	
Through Bookings—Steamboat prop'n	26,717	20,700	6,017	
Goods Freights	2,309	2,582		273
Mails	321	381		60
Charters	798	197	601	
Steward's Department	96		96	
Other Collections	259	246	13	
			7,323	333
	39,774	32,784	6,990	
Expenditure.				
Salaries and Wages	10,201	10,711		510
Agents' Wages	1,028	1,033		5
Repairs, Maintenance and Renewals	8,571	8,972		401
Coals, including Porterage	8,456	10,294		1,838
Oil, Waste and General Stores	1,192	1,519		327
Advertising, Printing and Stationery	1,095	1,372		277
Harbour and Light Dues	2,456	2,404	52	
General Expenses	1,412	967	445	
Claims and Damages	67	159		92
Charters and Boat Hires	335	184	151	
Steward's Department		591		591
	34,813	38,206	648	4,041

	1893 £	1892 £	Difference Increase £	Decrease £
Expenditure forward	34,813	38,206	648	4,041
Receipts forward	39,774	32,784		3,393
Balance to Net Revenue Account	Cr 4,961	Dr 5,422	10,383	

Net Revenue Account

	1893 £	1892 £	Difference Increase £	Decrease £
Insurance	2,157	2,353		196
Depreciation at 10%	13,045	13,045		
Interest on Balance of Advances	3,993	4,651		*658
Interest on Bank Balances	67	291		224
	19,262	20,340		1,078
Revenue Account				
1892 Balance Debit		5,422	} 10,383	
1893 Balance Credit	4,961			
	14,301	25,762	Net decrease in Traffic Debit £11,461	

* Note : This decreases as the balance is being written down.

ABSTRACT shewing a comparison in 1893 and 1892 as affecting the respective Boats

Steamer	1893				1892				Net improvement in each case, there being no decreases
	Receipts	Expend.	Balance Over	Under	Receipts	Expend	Balance Over	Under	
	£	£	£	£	£	£	£	£	£
Duchess of Hamilton	3,998	9,099		5,101	3,169	10,153		6,984	1,883
Galatea	6,181	7,544		1,363	5,027	8,352		3,325	1,962
Caledonia	6,491	6,117	374		5,108	6,281		1,173	1,547
Marchioness of Breadalbane	5,449	6,322		873	4,208	6,690		2,482	1,609
Marchioness of Bute	6,461	6,178	283		4,846	6,672		1,826	2,109
Meg Merrilies	3,574	6,262		2,688	3,011	6,600		3,589	901
Madge Wildfire	3,984	5,872		1,888	4,166	6,086		1,920	32
Marchioness of Lorne	3,928	6,906		2,978	3,269	7,440		4,171	1,193
			657	14,891					
	40,066	54,300		14,234	32,804	58,274		25,470	11,236

CALEDONIAN STEAM PACKET COMPANY, LTD

STATISTICAL DATA FOR THE YEAR 1892

Steamers	No of days run	No of miles run	Number of passengers carried: Railway	Local	Total	Aver. fare per pass	Aver. earnings per mile		Working expenses cost per mile, including repairs		Cost per day including repairs, insurance and interest		
						pence	s	d	s	d	£	s	d
Duchess of Hamilton	139	17,009	42,757	24,538	67,295	9·94	3	8¾	6	9¾	52	10	10
Galatea	234	22,492	128,979	34,728	163,707	6·86	4	5½	5	0	27	16	8
Caledonia	248	24,414	140,011	53,775	193,786	5·80	4	2	3	5¼	19	17	7
Marchioness of Breadalbane	229	23,404	123,082	33,647	156,729	6·07	3	7	3	9½	22	16	1½
Marchioness of Bute	230	20,961	137,421	20,497	157,918	6·79	4	7½	4	2½	22	12	2
Meg Merrilies	259	25,968	69,387	26,476	95,863	6·72	2	4	3	8¾	20	18	8¾
Madge Wildfire	265	27,298	57,419	29,795	87,214	8·16	3	0½	3	1¼	18	7	2
Marchioness of Lorne	276	19,687	73,557	30,093	103,650	7·25	3	4	4	9	20	16	4½
	1,880	181,233	772,613	253,549	1,026,162	6·87	3	7½	4	2½	205	15	7¾

STATISTICAL DATA FOR THE YEAR 1893

Steamers	No of days run	No of miles run	Number of passengers carried: Railway	Local	Total	Aver. fare per pass	Aver. earnings per mile		Working expenses cost per mile, including repairs		Cost per day including repairs, insurance and interest		
						pence	s	d	s	d	£	s	d
Duchess of Hamilton	134	16,550	46,236	22,072	68,308	11·00	4	10	5	11¾	46	13	8½
Galatea	182	21,072	141,441	44,126	185,567	7·15	5	10¼	4	7½	31	6	10¾
Caledonia	230	22,991	138,959	52,430	191,389	7·54	5	7¾	3	7½	20	16	1
Marchioness of Breadalbane	289	29,632	153,728	47,881	201,609	6·30	3	8	2	9¾	16	15	10¾
Marchioness of Bute	256	22,826	158,853	32,235	191,088	7·70	5	8	3	6	18	7	7
Meg Merrilies	246	23,681	87,602	28,126	115,728	6·91	2	11½	3	9½	20	8	5¼
Madge Wildfire	261	29,308	53,236	47,126	100,362	6·70	2	8	2	9¼	17	11	3
Marchioness of Lorne	276	19,280	68,840	30,183	99,025	9·14	3	11	4	3	18	5	11½
	1,874	185,340	848,897	304,179	1,153,076	7·49	4	3½	3	9	190	5	9¾

Appendix Ten

FLEET LISTS

The following lists include all vessels known to have sailed regularly on the Clyde during the period 1889-1914, with the exception of one or two ships used primarily as cargo carriers and some MacBrayne West Highland steamers which occasionally deputised on the Clyde. It should be noted that apart from building and breaking-up dates, all information relates to the various steamers only during the period 1889-1914.

The following abbreviations have been used :

C.D.	Compound Diagonal
cyl.	Cylinder(s)
D.D.	Double Diagonal
D.Osc.	Diagonal Oscillating
H.P.	High Pressure
I.P.	Intermediate Pressure
L.P.	Low Pressure
lb.	Pounds (per square inch)
N.B.	New Boiler
Osc.	Oscillating
P.S.	Paddle Steamer
S.D.	Single Diagonal
S.S.	(Single) Screw Steamer
St.	Steeple
T.C.D.	Tandem Compound Diagonal
T.E.	Triple Expansion
Tr.S.S.	Triple Screw Steamer

Type	Name	Built / Broken up	Builders / Engineers	Owners 1889–1914	Dimensions	Boilers	Machinery	Remarks
Iron P.S.	ADELA	1877 / ?	Caird & Co / Blackwood & Gordon	Capt. A. Campbell 1889–1890	207·7'×19·2'×7·4'	Haystack 50 lb	Ex *Lady Gertrude* S.D. 49"×54"	Sold off Clyde 1890
	ARDMORE (see *Sultan*)							
Iron P.S.	ARGYLE	1866 / ?	Barclay, Curle & Co / Do.	Capt. A. Campbell 1889–1890	177·3'×17·5'×7·6'	Haystack	Ex *Alma* St.	Sold off Clyde 1890
Steel Tr.S.S.	ATALANTA	1906 / 1947	John Brown & Co Ltd / Do.	G. & S.W. Ry	210·4'×30·1'×10·3'	2 Navy	Direct drive triple screw turbines – 1 H.P. and 2 L.P.	
Iron P.S.	ATHOLE	1866 / 1899	Barclay, Curle & Co / Do.	Bute S.P. Co Ltd 1889–98 Capt. John Williamson 1898–9	192·1'×18·5'×7·8'	Haystack	St. 48"×48"	
Iron P.S.	BALMORAL	1842 / 1891	J. Barr / Barr & McNab	Capt. Buchanan	136·7'×18·2'×7·9'		St. 47"×50"	Formerly *Lady Brisbane*
Iron P.S.	BENMORE	1876 / 1923	T. B. Seath & Co / W. King & Co	Capt. Buchanan 1889–92 Capt. John Williamson 1892–1914	201·2'×19·1'×7·3'	Haystack 50 lb	S.D. 50"×36"	
Steel P.S.	CALEDONIA	1889 / 1933	J. Reid & Co / Rankin & Blackmore	C.S.P. Co Ltd	200·4'×22·0'×7·5'	2 Navy 90 lb N.B. 1903	T.C.D. 30" & 54"×60"	
Steel P.S.	CARRICK CASTLE (see *Culzean Castle*)							
Steel P.S.	CULZEAN CASTLE	1891 / c 1931	Southampton Shipbuilding & Engineering Co	Glasgow, Ayrshire & Campbeltown Steamboat Co Ltd 1895–1900	244·6'×27·6'×10·3'		T.E. 26¼", 40" & 64¼"×60"	Formerly *Windsor Castle*. Sold off Clyde 1900
Steel P.S.	CHANCELLOR	1880 / ?	R. Chambers & Co / (1) M. Paul & Co (2) Blackwood & Gordon	Lochgoil & Lochlong Steamboat Co 1889–91 G. & S.W. Ry 1891–1901	199·7'×21·1'×8·2'	Haystack 50 lb N.B. 1892	Compounded 1892 Sold off Clyde 1901	
Steel P.S.	COLUMBA	1878 / 1936	J. & G. Thomson / Do.	David MacBrayne 1878–1905 David MacBrayne Ltd 1905–14	301·4'×27·1'×9·4'	(1) 4 Horizontal 50 lb (2) 2 Haystacks 55 lb	2 cyl. Osc. 53"×66"	N.B. 1900
Iron P.S.	CUMBRAE	1863 / 1892	Barclay, Curle & Co / J. Barr	Hill & Co	176·7'×17·6'×6·8'	Haystack	St. 54"×42"	
Steel P.S.	DANDIE DINMONT	1895 / 1936	A. & J. Inglis / Do.	N.B.S.P. Coy 1895–1902 N.B. Ry 1902–14	195·2'×22·1'×7·2' 209·6'×22·1'×7·2'	Haystack 50 lb	S.D. 48"×66"	Lengthened 1912

Type	Name	Built Broken up	Builders Engineers	Owners 1889–1914	Dimensions	Boilers	Machinery	Remarks
	DANIEL ADAMSON (see *Shandon*)							
Steel S.S.	DAVAAR	1885 1943	London & Glasgow Shipbuilding & Engineering Co Ltd	Campbeltown & Glasgow Steam Packet Joint Stock Co Ltd	217·8'×27·0'×12·9'	N.B. 1903	2 cyl compound 29" & 58"×42"	2 funnels prior to 1903
Iron & Steel P.S.	DIANA VERNON	1885 ?	Barclay, Curle & Co Do.	N.B.S.P. Coy 1889–1901	180·5'×18·1'×7·1'	Haystack 45 lb N.B. 1890	S.D. 43"×60"	Sold off Clyde 1901
Steel Tr.S.S.	DUCHESS OF ARGYLL	1906	Wm. Denny & Bros Denny & Co	C.S.P. Co Ltd	250·0'×30·1'×10·1'	Double ended	Direct drive triple screw turbines – 1 H.P. and 2 L.P.	Sold to Admiralty 1952
Steel P.S.	DUCHESS OF FIFE	1903 1953	The Fairfield Shipbuilding & Engineering Co Ltd	C.S.P. Co Ltd	210·3'×25·0'×8·5'	2 Navy	T.E. (4 cyl) (2) 16¼", 35" & 52"×54"	
Steel P.S.	DUCHESS OF HAMILTON	1890 1915	Wm Denny & Bros Denny & Co	C.S.P. Co Ltd	250·0'×30·1'×10·1'	3 Navy 120 lb N.B. 1906	C.D. 34¼" & 60"×60"	
Steel P.S.	DUCHESS OF MONTROSE	1902 1917	John Brown & Co Ltd Do.	C.S.P. Co Ltd	210·0'×25·1'×8·7'	2 Navy	T.E. (4 cyl) (2) 16¼", 35¼" & 53"×54"	
Steel P.S.	DUCHESS OF ROTHESAY	1895 1946	J. & G. Thomson Ltd Do.	C.S.P. Co Ltd	225·6'×26·1'×8·6'	Double ended 150 lb N.B. 1914	C.D. 27½" & 58"×54"	
	DUCHESS OF YORK (see *Jeanie Deans*)							
Iron P.S.	EAGLE	1864 1899	Charles Connel & Co W. King & Co	Capt. Buchanan 1889–94	219·5'×20·5'×7·3'	Haystack N.B. 1889	S.D. 50¼"×56"	Sold off Clyde 1894
Steel P.S.	EAGLE III	1910 1946	Napier & Miller Ltd A. & J. Inglis Ltd	Buchanan Steamers Ltd	215·0'×25·1'×8·1'	Haystack	S.D. 52"×72"	
Iron P.S.	EDINBURGH CASTLE	1879 1913	R. Duncan & Co Rankin & Blackmore	Lochgoil & Lochlong Steamboat Co 1889–1895 Lochgoil & Lochlong Steamboat Co Ltd 1895–1909 Lochgoil & Inveraray Steamboat Co Ltd 1909–12 Turbine Steamers Ltd 1912–13	205·3'×19·9'×7·6'	Haystack 50 lb N.B. 1903	S.D. 50"×66"	

	Name	Dates	Builder	Owners	Dimensions	Boiler	Engine	Notes
Iron P.S.	ELAINE	1867 1908	R. Duncan & Co Rankin & Blackmore	Capt. Buchanan 1889–1905 Buchanan Steamers Ltd 1905–8	175·0' × 17·1' × 6·6'	Haystack 40 lb	2 cyl Osc. 28" × 44"	Sold off Clyde 1906
Steel P.S.	GALATEA	1889 1913	Caird & Co Do.	C.S.P. Co Ltd 1889–1906	230·1' × 25·1' × 7·8'	4 Navy 109 lb	C.D. 34" & 64" × 72"	
Iron P.S.	GARELOCH	1872 1906	Henry Murray & Co D. Rowan & Co	N.B.S.P. Coy 1889–91	180·0' × 18·2' × 6·8'	Haystack 40 lb	2 cyl Osc. 35" × 54"	To Galloway Saloon Steam Packet Co Ltd Leith 1891
Steel P.S.	GLEN ROSA	1893 1939	J. & G. Thomson Ltd Do.	G. & S.W. Ry	200·0' × 25·0' × 8·3'	Double ended 150 lb	C.D. 26" & 55" × 54"	
Steel P.S.	GLEN SANNOX	1892 1925	J. & G. Thomson Ltd Do.	G. & S.W. Ry	260·5' × 30·1' × 10·1'	1 Double ended & 1 Navy 150 lb	C.D. 34¾" & 74" × 60"	
Steel P.S.	GLENMORE	1895 ?	Russell & Co Rankin & Blackmore	Capt. John Williamson 1895–6	190·3' × 21·1' × 7·2'	1 Navy 120 lb	C.D. 22" & 44" × 51"	Sold to Russia 1896
Steel P.S.	GRENADIER	1885 1928	J. & G. Thomson Do.	David MacBrayne 1889–1905 David MacBrayne Ltd 1905–14	222·9' × 23·1' × 9·3'	2 Scotch replaced in 1902 by 2 Haystacks (95 lb)	Compound Osc. 2 cyl 30" & 58" × 51"	
Iron P.S.	GUINEVERE	1869 1892	R. Duncan & Co Rankin & Blackmore	Capt. Buchanan 1889–92	200·3' × 19·1' × 6·8'	2 Haystacks	2 cyl Osc. 36" × 48"	Sold off Clyde and lost at sea 1892
Iron P.S.	GUY MANNERING	1877 1913	Caird & Co Do.	N.B.S.P. Coy 1889–94 Capt. Buchanan 1894–1905 Buchanan Steamers Ltd 1905–12	205·5' × 20·0' × 7·7'	Haystack 50 lb N.B. 1891	S.D. 50" × 72"	Formerly *Sheila* Latterly *Isle of Bute* Sold off Clyde 1912
Iron P.S.	HERALD	1866 ?			221·5' × 22·0' × 10·3'		2 cyl Osc. 44" × 57"	
Iron P.S.	IONA	1864 1936	J. & G. Thomson Do.	David MacBrayne 1889–1905 David MacBrayne Ltd 1905–14	255·5' × 25·6' × 9·0'	Horizontals replaced by 2 Haystacks in 1891	2 cyl Osc. 50¾" × 51"	
Steel P.S.	ISLE OF ARRAN	1892 1936	T. B. Seath & Co W. King & Co	Capt. Buchanan 1892–1905 Buchanan Steamers Ltd 1905–14	210·0' × 24·1' × 7·4'	Haystack 60 lb	S.D. 52" × 60"	
	ISLE OF CUMBRAE (see *Jennie Deans*)							
	ISLE OF SKYE (see *Madge Wildfire*)							

Type	Name	Built Broken-up	Builders Engineers	Owners 1889–1914	Dimensions	Boilers	Machinery	Remarks
Iron P.S.	IVANHOE	1880 1920	D. & W. Henderson & Co Do.	Frith of Clyde S.P. Co Ltd 1889–1897 C.S.P. Co Ltd 1897–1911 Firth of Clyde S.P. Co Ltd 1911–14 Turbine Steamers Ltd 1914	225·3′ × 22·2′ × 8·3′	2 Haystacks 50 lb	2 cyl D osc 43″ × 66″	
Iron & Steel P.S.	JEANIE DEANS	1884 1920	Barclay, Curle & Co Do.	N.B.S.P. Coy 1889–1896 W. Dawson Reid 1898–1904 Capt. Buchanan 1904–5 Buchanan Steamers Ltd 1905–14	210·0′ × 20·1′ × 7·6′	Haystack 45 lb N.B. 1901	S.D. 50″ × 72″	Sold off Clyde 1896. Returned 1898. Renamed *Duchess of York* 1898. Renamed *Isle of Cumbrae* 1904
Steel P.S.	JUNO	1898 1932	Clydebank Engineering & Shipbuilding Co Ltd Do.	G. & S.W. Ry	245·0′ × 29·1′ × 9·7′	Double ended 150 lb	C.D. 32″ & 71″ × 60″	
Steel P.S.	JUPITER	1896 1935	J. & G. Thomson Ltd Do.	G. & S.W. Ry	230·0′ × 28·1′ × 9·0′	Double ended 150 lb	C.D. 30½″ & 65″ × 60″	
Steel P.S.	KENILWORTH	1898 1938	A. & J. Inglis Do.	N.B.S.P. Coy 1898–1902 N.B. Ry 1902–14	215·0′ × 23·1′ × 7·6′	Haystack 65 lb	S.D. 52″ × 72″	
Steel Tr.S.S.	KING EDWARD	1901 1951	Wm. Denny & Bros Denny & Co (Boiler) Parsons Marine Steam Turbine Co Ltd (Engines)	Turbine Syndicate 1901–2 Turbine Steamers Ltd 1902–14	250·5′ × 30·1′ × 10·0′	Double ended 150 lb	Direct drive triple screw steam turbines – 1 H.P. & 2 L.P.	Originally with 5 propellers
Iron S.S.	KINLOCH	1878 1926	A. & J. Inglis Do.	Campbeltown & Glasgow Steam Packet Joint Stock Co Ltd	205·0′ × 24·1′ × 12·7′	N.B. 1890 and 1914	2 cyl compound 29″ & 54″ × 42″	
Iron S.S.	KINTYRE	1868 1907	Robertson & Co Kincaid, Donald & Co	Campbeltown & Glasgow Steam Packet Joint Stock Co Ltd 1889–1907	184·7′ × 22·9′ × 11·5′	N.B. 1893	2 cyl compound 26″ & 48″ × 30″	Sunk in collision on Clyde 1907
	KYLEMORE (see *Vulcan*)							
Steel P.S.	LADY CLARE	1891 1928	J. McArthur & Co Hutson & Corbett	N.B.S.P. Coy 1891–1902 N.B. Ry 1902–6	180·5′ × 19·3′ × 6·5′	Haystack 55 lb	S.D. 44″ × 60″	Sold off Clyde 1906

	Name	Built / Broken up	Builder	Owners	Dimensions	Boiler	Engine	Notes
	LADY OF THE ISLES (see Lord of the Isles (I))							
Steel P.S.	LADY ROWENA	1891 / 1922	S. McKnight & Co / Hutson & Corbett	N.B.S.P. Coy 1891–1902 / N.B. Ry 1902–3 / Capt. A. W. Cameron 1911–14	200·5′ × 21·1′ × 7·2′	Haystack 55 lb N.B. 1901	S.D. 50″ × 72″	Sold off Clyde in 1903; returned in 1911
Iron P.S.	LANCELOT	1868 / ?	R. Duncan & Co / Rankin & Blackmore	Capt. A. Campbell 1889–90	191·2′ × 18·0′ × 6·9′	Haystack 40 lb	2 cyl osc 32¼″ × 48″	Sold off Clyde 1890
Iron P.S.	LORD OF THE ISLES (I)	1877 / 1904	D. & W. Henderson & Co / Do.	Glasgow & Inveraray Steamboat Coy 1889–90 / Glasgow & Inveraray Steamboat Co Ltd 1890–91 / W. Dawson Reid 1904	246·0′ × 24·2′ × 9·0′	2 Haystacks 50 lb	2 cyl diag osc 46″ × 66″	Away from Clyde 1891–1904 Latterly Lady of the Isles (1904)
Steel P.S.	LORD OF THE ISLES (II)	1891 / 1928	D. & W. Henderson & Co / Do.	Glasgow & Inveraray Steamboat Co Ltd 1891–1909 / Lochgoil & Inveraray Steamboat Co Ltd 1909–12 / Turbine Steamers Ltd 1912–14	255·0′ × 25·6′ × 9·1′	2 Haystacks 50 lb N.B. 1908	2 cyl diag osc 48″ × 66″	
Steel P.S.	LUCY ASHTON	1888 / 1949	T. B. Seath & Co / (1) Hutson & Corbett / (2) A. & J. Inglis (1902)	N.B.S.P. Coy 1888–1902 / N.B. Ry 1902–14	190·0′ × 21·1′ × 7·2′	(1) Haystack 50 lb / (2) Haystack 110 lb (N.B. 1902)	(1) S.D. 52″ × 60″ / (2) C.D. 28″ & 52″ × 60″	
Steel P.S.	MADGE WILDFIRE	1886 / 1945	S. McKnight & Co / Hutson & Corbett	C.S.P. Co Ltd 1889–1911 / Capt. A. W. Cameron 1911–13 / Buchanan Steamers Ltd 1913–14	190·0′ × 20·0′ × 7·1′	Haystack 50 lb 1889–1891 / 2 Navy 1891–1914 (N.B. 1903)	(1) S.D. 49″ × 60″ / Rebuilt as (2) T.C.D. 27″ & 49″ × 60″ (1891)	Renamed Isle of Skye in 1913
Steel P.S.	MARCHIONESS OF BREADALBANE	1890 / 1935	J. Reid & Co / Rankin & Blackmore	C.S.P. Co Ltd	200·4′ × 22·1′ × 7·5′	2 Navy 100 lb N.B. 1901	T.C.D. 30″ & 54″ × 60″	
Steel P.S.	MARCHIONESS OF BUTE	1890 / 1923	J. Reid & Co / Rankin & Blackmore	C.S.P. Co Ltd 1889–1908	200·4′ × 22·1′ × 7·5′	2 Navy 100 lb N.B. 1901	T.C.D. 30″ & 54″ × 60″	Sold off Clyde 1908
Steel P.S.	MARCHIONESS OF LORNE	1891 / 1923	Russell & Co / Rankin & Blackmore	C.S.P. Co Ltd	200·0′ × 24·0′ × 8·3′	2 Navy 140 lb N.B. 1897	T.E. (4cyl) (2) 17½″ 30″ & 49″ × 60″	
Steel P.S.	MARMION	1906 / 1941	A. & J. Inglis Ltd / Do.	N.B. Ry	210·0′ × 24·0′ × 8·3′	Haystack	C.D. 31″ & 56″ × 66″	

Type	Name	Built Broken-up	Builders Engineers	Owners 1889–1914	Dimensions	Boilers	Machinery	Remarks
Iron P.S.	MARQUIS OF BUTE	1868 1908	Barclay, Curle & Co Do.	A. Williamson, Sr 1889–91 G. & S.W. Ry 1891–1904	196·6′ × 18·1′ × 7·3′	Haystack 45 lb	S.D. 48″ × 60″	Sold off Clyde 1904
Steel P.S.	MARS	1902 1918	John Brown & Co Ltd Do	G. & S.W. Ry	200·4′ × 26·1′ × 8·6′	2 Navy	C.D. 28¼″ & 53″ × 54″	
Iron P.S.	MEG MERRILIES	1883 1921	Barclay, Curle & Co Do.	C.S.P. Co Ltd 1889–1902	210·3′ × 21·4′ × 7·2′	Haystack 1889–97 (N.B. 1892) 2 Navy 1897–8 2 Haythorn Water tube 1898–1900 2 Navy from 1900	D.D. 43″ × 60″ Converted in 1898 to C.D. 24″ & 43″ × 60″	Sold off Clyde 1902
Steel P.S.	MERCURY	1892 1933	Napier, Shanks & Bell D. Rowan & Son	G. & S.W. Ry	220·5′ × 26·0′ × 9·2′	2 Navy 115 lb N.B. 1912	C.D. 33″ & 62″ × 60″	
Steel P.S.	MINERVA	1893 1917	J. & G. Thomson Ltd Do.	G. & S.W. Ry	200·0′ × 25·0′ × 8·3′	Double ended 150 lb N.B. 1902	C.D. 26″ & 55″ × 54″	
Steel P.S.	NEPTUNE	1892 1917	Napier, Shanks & Bell D. Rowan & Son	G. & S.W. Ry	220·5′ × 26·0′ × 9·2′	2 Navy 115 lb N.B. 1912	C.D. 33″ & 62″ × 60″	
Steel Tr.S.S.	QUEEN ALEXANDRA (I)	1902 1937	Wm. Denny & Bros Denny & Co (Boiler) The Parsons Marine Steam Turbine Co Ltd (Engines)	Capt. John Williamson 1902 Turbine Steamers Ltd 1902–11	270·0′ × 32·1′ × 11·0′	Double ended 150 lb	Direct drive triple screw steam turbines 1 H.P. and 2 L.P.	Originally had 5 propellers Sold off Clyde 1911
Steel Tr.S.S.	QUEEN ALEXANDRA (II)	1912 1958	Wm. Denny & Bros Denny & Co	Turbine Steamers Ltd	270·3′ × 32·1′ × 11·0′	Double ended	Direct drive triple screw steam turbines 1 H.P. and 2 L.P.	
Steel P.S.	QUEEN EMPRESS	1912 1946	Murdoch & Murray Rankin & Blackmore	Capt. John Williamson	210·0′ × 25·6′ × 8·4′	2 Navy	C.D. 27⅞″ & 54″ × 54″	
Steel P.S.	REDGAUNTLET	1895 c.1934	Barclay, Curle & Co Ltd Do.	N.B.S.P.Co 1889–1902 N.B. Ry 1902–09	215·0′ × 22·1′ × 7·4′	Haystack 60 lb	S.D. 53″ × 72″	To Galloway Saloon Steam Packet Co Ltd Leith 1909
Iron P.S.	SCOTIA	1880 c.1914	H. McIntyre & Co W. King & Co	Capt. Buchanan 1889–91 G. & S.W. Ry 1891–3	211·2′ × 21·8′ × 8·3′	2 Haystacks 50 lb N.B. 1892	2 cyl St. 45″ × 48″	Sold off Clyde 1893
Iron P.S.	SHANDON	1864 1895	Blackwood & Gordon Do.	Capt. Buchanan 1889–94	163·2′ × 18·7′ × 7·0′	Haystack	D.D. 32″ × 51″	To Manchester Ship Canal 1894 Brief return to Clyde 1895
Steel P.S.	STRATHMORE	1897 c.1941	Russell & Co Rankin & Blackmore	Capt. John Williamson 1897–1908	200·5′ × 24·1′ × 7·7′	1 Navy 120 lb	C.D. 23⅜″ & 47″ × 51″	Sold off Clyde 1908

Type	Name	Built / Broken up	Builders / Engineers	Owners	Dimensions	Boiler	Engines	Remarks
Iron P.S.	SULTAN	1861 1919	Robertson & Co W. King & Co	A. Williamson, Sr 1889–91 G. & S.W. Ry 1891–3 Capt. John Williamson 1893–4	176·0′ × 16·6′ × 7·2′	Haystack 30 lb	St. 45″ × 42″	Latterly *Ardmore* Sold to D. MacBrayne as his *Gairlochy* 1894 To Caledonian Canal
Iron P.S.	SULTANA	1868 c.1907	Robertson & Co W. King & Co	A. Williamson, Sr 1889–91 G. & S.W. Ry 1891–6 Capt. John Williamson 1896–9 Lochfyne & Glasgow S.P. Co Ltd 1899–1900	188·1′ × 18·3′ × 7·3′	Haystack 45 lb	S.D. 49″ × 54″	Sold off Clyde 1900
Steel P.S.	TALISMAN	1896 1934	A. & J. Inglis Do.	N.B.S.P. Coy 1896–1902 N.B. Ry 1902–14	215·0′ × 23·0′ × 7·5′	Haystack 65 lb N.B. 1910	S.D. 52″ × 72″	
Iron P.S.	VICEROY	1875 1911	D. & W. Henderson & Co Hutson & Corbett	A. Williamson, Sr 1889–91 G. & S.W. Ry 1891–1907	194·7′ × 20·1′ × 7·1′ 208·9′ × 20·1′ × 7·1′	Haystack 50 lb N.B. 1897	S.D. 51¼″ × 60″	Lengthened 1891 Sold off Clyde 1907
Steel P.S.	VICTORIA	1886 ?	Blackwood & Gordon Do.	Capt. A. Campbell 1889–90 Scottish Excursion Steamer Co Ltd 1892–3 The Clyde Steamers Ltd 1897	222·4′ × 23·1′ × 8·0′	2 Haystacks (N.B. 1892)	D.D. 40″ × 66″	Sold off Clyde in 1890–1 and in 1894–6 Finally sold off Clyde 1897
Iron P.S.	VIVID	1864 1902	Barclay, Curle & Co Do.	Capt. Buchanan 1889–1902	197·3′ × 18·2′ × 7·8′	Haystack	St. 48″ × 48″	
Steel P.S.	VULCAN	1897 1940	Russell & Co Rankin & Blackmore	G. & S.W. Ry 1904–8 Capt. John Williamson 1908–14	200·5′ × 24·1′ × 7·7′	1 Navy 120 lb	C.D. 23¾″ & 47″ × 51″	Originally *Kylemore* then *Britannia* Latterly *Kylemore* again
Steel P.S.	WAVERLEY	1899 1940	A. & J. Inglis Do.	N.B.S.P. Coy 1899–1902 N.B. Ry 1902–14	235·0′ × 26·1′ × 8·4′	Haystack 110 lb	C.D. 37″ & 67″ × 66″	
Iron P.S.	WINDSOR CASTLE	1875 1908	T. B. Seath & Co W. King & Co	Lochgoil & Lochlong Steamboat Co 1889–95 Lochgoil & Lochlong Steamboat Co Ltd 1895–1900	195·8′ × 19·0′ × 7·2′	Haystack 50 lb	S.D. 50″ × 54″	Sold off Clyde 1900

FLAG COLOUR CODE

 blue

red

green

yellow

white

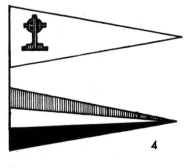

4

1

5

MARS

2

6

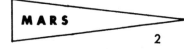

3

7

8

1. Captain Buchanan, and Buchanan Steamers Ltd
2. Steamer name pennant (Red lettering)
3. Wemyss Bay Company (Captain A. Campbell). Two flags
4. Campbeltown & Glasgow Steam Packet Joint Stock Co Ltd (Two flags)
5. Glasgow & Inveraray Steamboat Co Ltd
6. Glasgow & South Western Railway
7. Glasgow, Ayrshire & Campbeltown Steamboat Co Ltd
8. Captain A. Cameron

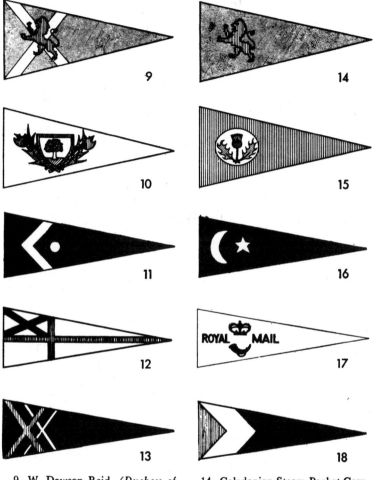

9. W. Dawson Reid. *(Duchess of York)*
10. Firth of Clyde Steam Packet Co Ltd *(Ivanhoe* from 1911)
11. P. & A. Campbell Ltd
12. Lochgoil & Lochlong Steamboat Co Ltd
13. David MacBrayne, and David MacBrayne Ltd
14. Caledonian Steam Packet Company Ltd
15. North British Railway
16. The 'Turkish' fleet; Captain John Williamson; Turbine Syndicate, and Turbine Steamers Ltd
17. Royal Mail pennant (Red lettering, etc)
18. Frith of Clyde Steam Packet Co Ltd *(Ivanhoe* to 1897)

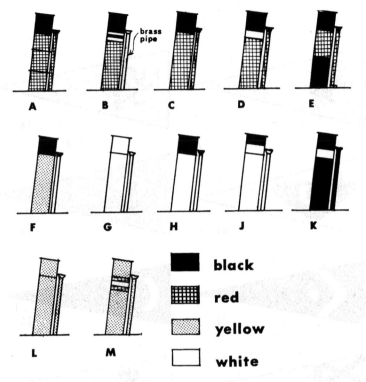

FUNNEL COLOUR CODE

KEY TO FUNNEL COLOURS

A David MacBrayne, until about 1906.
B Glasgow & Inveraray Steamboat Co Ltd; Lochgoil & Lochlong Steamboat Co Ltd: Lochgoil & Inveraray Steamboat Co Ltd: and Turbine Steamers Ltd (for the *Lord of the Isles* and *Edinburgh Castle* only while in their ownership).
C Glasgow & South Western Railway: David MacBrayne, Ltd.: Captain A. Cameron: Glasgow, Ayrshire & Campbeltown Steamboat Co Ltd.
D North British Steam Packet Company; North British Railway.
E Campbeltown & Glasgow Steam Packet Joint Stock Co Ltd.
F Scottish Excursion Steamer Co Ltd.: the Clyde Steamers Ltd.: Captain A. Cameron; *Culzean Castle* also bore these colours for a time.
G P. & A. Campbell, Ltd.: and Firth of Clyde Steam Packet Co Ltd at first.
H Wemyss Bay Company (Captain A. Campbell): Captain John Williamson, from June, 1898: Turbine Syndicate; and Turbine Steamers, Ltd.
J Firth of Clyde Steam Packet Co Ltd (later stage).
K Captain Buchanan, and Buchanan Steamers, Ltd.: the 'Turkish' fleet: of Captain Alex. Williamson, Sr.: Captain John Williamson, until June, 1898.
L Frith of Clyde Steam Packet Co Ltd *(Ivanhoe)*, and the Caledonian Steam Packet Co Ltd. In the former case, the top stay ring was painted black, but this feature could not be detected at any distance.
M Dawson Reid's colours for the *Duchess of York*.

AUTHOR'S NOTES
AND ACKNOWLEDGMENTS

In writing this book I have been fortunate in having had access to hitherto unpublished records and I am indebted to Mr Robert M. Hogg, Custodian of Records, British Rail, Edinburgh, for his kindness in making available to me the minute books and other records of the Scottish railways and steam packet companies, and granting permission to quote freely from these sources. Mr Hogg's services to the cause of Scottish transport history have long been recognised by other writers and it is with pleasure that I now associate myself with them in thanking Mr Hogg and his staff for their interest in this project over a period of several years.

I have also to record my thanks and appreciation in respect of permission granted by Mr A. W. H. Pearsall, Curator of the National Maritime Museum, Greenwich, to examine and quote from the records of Denny & Co and Wm Denny & Bros, Dumbarton, and in particular to reproduce official photographs of the early turbine steamers. For permission to use other Denny records and to reproduce as an appendix the original Turbine Syndicate agreement, my thanks are due to Dr P. L. Payne, Colquhoun Lecturer in Business History at the University of Glasgow.

I acknowledge with pleasure the generous assistance of Messrs G. E. Langmuir, G. M. Stromier, L. J. Vogt and K. V. Norrish in allowing me to use certain photographs from their collections, while the editor of *Engineering* has kindly permitted reproduction of an illustration of the machinery of the paddle steamer *Caledonia*. To Mr Charles Black, City Librarian, Glasgow, his colleague Mr A. Thomson, and staff of the Mitchell Library, Glasgow, I am much indebted for valuable assistance over a long period, chiefly in the field of newspaper research.

I should also like to record my appreciation of the help of various gentlemen who have provided information; Mr H. G. Owen, of Swansea, has been of great assistance in tracing details of the *Britannia*'s voyage to the Clyde in 1901, while Mr Frank Beken, of Cowes, has given me the benefit of his long experience of yachting history which I have been glad to use in the preparation of notes on Clyde racing. I have also received help and advice from Mr

W. M. Martin, Librarian of the Royal Burgh of Dumbarton, Mr R. J. G. Easthope (Denny matters), and Mr Graham King (machinery details), while Messrs G. E. Langmuir, G. M. Stromier, L. J. Vogt, G. C. Train and Dr J. A. Weir have been good enough to comment and give advice and suggestions. The proprietors of *The Ardrossan & Saltcoats Herald* and *The Oban Times* have given me facilities, for which I am grateful. Mr Alisdair Macdonald has gone to much trouble to prepare the drawings which adorn the book, and to him I am indebted for having undertaken this part of the work, and Mr Roy Hamilton takes credit for having discovered and brought to my notice the splendid Glasgow & South Western Railway poster which forms the basis of the frontispiece. Not least of all, I have great pleasure in thanking Mr John Thomas for having encouraged me in the first instance to undertake this work, and for his invaluable advice and help at various stages.

It would be ungenerous of me to conclude without making reference to three Clyde steamer enthusiasts, now dead, to whose kindness in past years I owe much. The Rev Wm C. Galbraith's encyclopaedic knowledge is still a byword; Mr Cameron Somerville's humour and ability to recall vividly the atmosphere of former years have been caught in a memorable booklet on which I have tried, however inadequately, to model this work; and to Mr Ronald B. McKim, more than to any other, was due my youthful enthusiasm for the technicalities of boilers and engines. These men knew and loved the Clyde and its steamers in their prime and in acknowledging my debt of gratitude to them it is my hope that this book will in some measure keep alive the memories which they treasured and shared with a younger generation.

A.J.S.P.

BIBLIOGRAPHY

BOOKS AND BOOKLETS

Acworth, W. J. *The Railways of Scotland*, 1890.

Duckworth, C. L. D. and Langmuir, G. E. *West Highland Steamers*. Richard Tilling, 1936. Revised and enlarged editions, 1950 and 1967.

Duckworth, C. L. D. and Langmuir, G. E. *Clyde River and Other Steamers*. Brown Son and Ferguson, 1938. Revised 1946 and 1969.

Galbraith, Rev Wm. C. *Sixty Years of the Lucy Ashton*. Clyde River Steamer Club, 1948.

Galbraith, Rev Wm. C. *Sixtieth Anniversary of the Caledonian Steam Packet Company, &c*. Clyde River Steamer Club, 1949.

McQueen, A. *Clyde River Steamers of the Last Fifty Years*. Gowans & Gray, Ltd, 1923.

McQueen, A. *Echoes of Old Clyde Paddle Wheels*. Gowans & Gray, Ltd, 1924.

Somerville, Cameron. *Colour on the Clyde*. The Buteman, 1959.

Stromier, G. M. *Steamers of the Clyde*. Nicholson, 1967.

Williamson, J. *The Clyde Passenger Steamer*. MacLehose, 1890.

SOURCE MATERIAL

Denny Collection—Engineering Records. National Maritime Museum, Greenwich

Denny Collection—Business Records. Department of Economic History, University of Glasgow

Minute Book—Caledonian Steam Packet Co Ltd, 1889-1914
Glasgow & South Western Railway, 1891-1914
North British Steam Packet Company, 1898-1902
North British Railway, 1902-1914

Timetables, public notices, financial records, and miscellaneous travel literature of the period. 1889-1914

TECHNICAL PAPERS

'The Marine Steam Turbine and its Application to Fast Vessels.' Paper read to the Institution of Engineers and Shipbuilders in Scotland by the Hon C. A. Parsons. 1901.

'The Steam Turbine and its Application to the Propulsion of Vessels.' Paper read to the Institution of Naval Architects by the Hon C. A. Parsons. 1903.

'The History and Development of Machinery for Paddle Steamers.' Paper read to the Institution of Engineers and Shipbuilders in Scotland by G. E. Barr. 1951.

TECHNICAL AND OTHER JOURNALS

The Engineer
Engineering
The Marine Engineer
Marine Engineering
The Railway Magazine
The Shipbuilder

NEWSPAPERS AND PERIODICALS

The Ardrossan & Saltcoats Herald
The Bailie
The Buteman
The Glasgow Herald
The Greenock Telegraph
The Dumbarton Herald & Lennox Herald
The Oban Times; and several other Clyde coast newspapers of the period.

MISCELLANEOUS

The Wotherspoon Collection. The Mitchell Library, Glasgow.
Records of the Clyde River Steamer Club, Glasgow.

INDEX

Illustrations are indicated by italic type, and fleet list references by bracketed page numbers.

Accidents and collisions, 52, 56-7, 64, 80, 98, 110, 115, 128-9, 139, 141, 149-50, 160, 163, 214, 217, 223-4, 226
Acworth, W. J., 212
Admiralty, 154-5, 163, 205, 229
Advertisements (Cal Ry), 190
Alley & McLellan's steering gear, 105, 123
American Civil War, 19
America's Cup, 226
Angeletti, Mr, 138
Angus, Mr, 103
Armstrong, Whitworth & Co, 155
Arran Express, The (Cal Ry), 51
Austin, Mr, 138

Balfour Browne, Mr, QC, 75
Barclay, Curle & Co, 28, 65, 119
Barnard, Robert, 163
Barr, Capt William, 25, 186, 245, 252
Belfast & County Down Ry, 87, 121, 129
Bell, Capt Duncan, 93
Berliner Philharmonisches Blas-Orchester, 229-30
Birmingham Railway Carriage & Wagon Co Ltd, 82
Blackwood & Gordon, 29, 111, 146
Board of Trade, 129, 135-7, 142, 150, 196, 199-200, 202-3, 214, 266, 269
Boilers—double ended, *240*, 247; haystack, *235*, 236-7, 269; Navy, 237, *238*; 239; Scotch, 247
Bournemouth, Swanage & Poole Steam Packet Co, 127
Bow rudder (Duchess of Argyll), 171
Brescian family—see Hayward
Bristol Channel, 31-2, 37-8, 77, 88, 148, 227
Britannia cruise to Clyde, 227-8
Brock, Capt, 110
Brock, Walter, 239
Broomielaw trade, 15, 21, 58, 117, 206-11
Brown, Mr, 84
Brown, John, & Co Ltd, 168-9, 172-3, 186-7
Buchanan, Capt William, and family, 29-32, 47, 52, 58, 67, 74, 95-6, 106, 110, 116, 119, 200, 213, 228, 256, 284, 286
Buchanan Steamers, Ltd, 201, 207, 210, 237, 252, 256, 258, 284, 286
Buchanan's Trustees, 74-5
Bute Steam Packet Co, 16

Caird & Co, 28, 37, 59, 115, 140
Caldwell, James, MP, 69-70
Caledonian Canal, 115
Caledonian Railway, 19, 22, 28, 31, 33-4, 37-9, 43, 46-7, 52, 55, 57-9, 65, 67-8, 73-4, 84-5, 95-6, 98-9, 118, 123, 167, 175-6, 184, 190, 203, 213, 236, 270
Caledonian Railway (Steam Vessels) Bill, 37, 39-40
Caledonian Steam Packet Co Ltd, 41, 44, 46-7, 56, 58-9, 65-6, 73, 75-6, 80, 99, 100, 116, 120, 132-3, 142, 146, 152, 167-8, 173, 181, 185, 190, 194-6, 205, 209, 220-1, 225-6, 228-9, 239, 241-3, 247, 251-4, 258, 272-5, 285-6
Callander & Oban Ry, 24
Cameron, Capt A. W., 209-10, 258, 285-6
Campbell, Capt Alex (Kilmun), 31, 37-8, 148, 227
Campbell, Capt Alexander (Wemyss Bay), 20, 32, 52, 55, 97-9, 111
Campbell, Capt Angus, *197*, 220
Campbell, P. & A., 236, 285-6
Campbell, Capt Peter, 31, 37-8, 57, 148 227
Campbell, Capt Robert, 29, 31-2, 38
Campbeltown & Glasgow Steam Packet Joint Stock Co Ltd, 30, 68, 112-3, 115, 117, 127, 192, 204, 257-8, 284, 286
Campbeltown & Machrihanish Light Ry, 181
Canadian Pacific Ry, 177
Cayzer, Mr, 37-8
Channel Fleet, 94, 160
Chetwynd, Miss A. M., 178
Clark, D. T., 127
Clark, Kenneth, 228
Clark, Malcolm T., 25, 40, 101, 206
Clyde Fortnight, 63, 97, 224-6
Clyde Salvage Co, 150
Clyde Shipbuilding & Engineering Co, 185, 192
Clyde Steamers, Ltd, 135, 137, 286
Clyde Steamship Owners' Association, 39, 68
Clydebank Engineering & Shipbuilding Co Ltd, 139-40
Cockburn, Councillor, 202
Cockburn, Mr, 174
Collisions—see Accidents & Collisions
Comrie, Mr, 73
Confederacy (Confederate States of America), 19
Constant, Mr, 151
Cooper David, 202
Cowan, John, 43
Crawford, Councillor, 70
Crosbie, Commissioner, 136
Cunard Steamship Co, 255
Custom House Quay, 16, 22, 46
Cut-off route (NBR), 133

Dalreoch Tunnel, 132
Darling, Robert, 27, 95, 118, 120, 147-8
Darling, The Misses, 103
Dawson Reid, Andrew, 135, 137, 139, 146, 192-3, 229, 258, 284, 286
Denny, Archibald, 155-6, 264
Denny, James, 165
Denny, Col John, 177
Denny, Peter, 159
Denny & Brace spark arrester, 222
Denny & Co, 165

Denny, William, & Bros, 48-50, 59-60, 145, 155-7, 164, 168-70, 173, 177, 181-2, 204, 239, 241, 252, 262-5
Derry & Moville Steam Packet Co, 131
Diamond Jubilee of Caledonian Ry, 204
Downie, Capt Donald, 105, 220
Drummond, Dugald, 43, 51
Drunkenness, 26-7
Dübs & Co, 51
Duchess of Fife—see Princess Louise
Duchess of Hamilton, 51
Duchess of Montrose, 186
Duckworth & Langmuir, 255
Duke of Argyll, 59-60
Duke of Fife, 183
Duke of Hamilton, 74, 170, 232
Dunoon Commissioners, 135-8

Earl of Dunraven, 225
Earl of Eglinton, 51
Earl of Rosse, 153
Edinburgh International Exhibition of 1886, 51
Edwards, Frederick, 86
Electric Light in NBR trains, 97
Engines—compound diagonal, 161, 179, 239, 241, 243; compound oscillating, 244; diagonal oscillating, 245; double diagonal, 239; double steeple, 245; screw, 245; simple oscillating, 244; single diagonal, 162, 235-6; steeple, 244-5; tandem compound, 161, 237; triple expansion, 179, 241-3
Engine details—Brock valve gear, 239; control platforms, 247-8; crankshafts, 249, 269; decoration, 249; elm paddle floats, 247, 269; entablatures, 248-9; impulse valves, Duchess of Fife, 242-3; paddle-wheels, 242, 245, 246, 247, 269; steel paddle floats, 247; Stephenson link valve gear, 241
Evening cruise, Turbine, 160

Fairfield Shipbuilding & Engineering Co Ltd, 48, 59, 190-1
Farmer & Stewart, 222
Federal Government of United States, 19
Ferry service, Greenock-Craigendoran, 120
Fife, William, 225
Firth of Forth, 101, 206
Firth of Clyde Steam Packet Co Ltd, 209, 211, 258, 285-6
Fleming & Ferguson, 59
Fletcher, Alfred E., 220
Fog, 222-3
Forced draught, 239
Forth Bridge, 22
Frith of Clyde Steam Packet Co Ltd, 26-7, 44, 133-4, 257, 285-6

Galloway Saloon Steam Packet Co, 101, 206
Gannaway's ventilation system, 105
German Bands, 228
Gilchrist, L. H., 148
Gillies, Capt, 20
Gillies & Campbell, 28
Glasgow, Ayrshire & Campbeltown Steamboat Co Ltd, 127, 284, 286
Glasgow, Barrhead & Kilmarnock Joint Ry, 47, 74
Glasgow International Exhibition of 1888, 51
Glasgow International Exhibition of 1901, 159, 183-4, 219, 227

Glasgow & Inveraray Steamboat Co Ltd, 25, 36-7, 40, 68, 95, 100-1, 104, 166, 206, 255, 284, 286
Glasgow, Paisley & Greenock Ry, 16, 19
Glasgow & South Western Ry, 21-2, 29, 31, 40, 43, 46-7, 49, 52, 57-8, 66-7, 73-4, 76-7, 79, 81, 83-5, 91-2, 94-5, 101, 109, 113-16, 118, 121, 123, 128-9, 132, 139-40, 146, 152, 156-7, 160, 163-4, 167-8, 172-5, 182, 184, 187-8, 192-6, 202, 205-6, 213-14, 217, 220, 228, 239, 241-3, 247, 249, 251-2, 254, 256, 258, 270, 284, 286
Glasgow & South Western Ry (Steam Vessels), Bill, 68, 73-5
Glasgow Steamers, Ltd, 146
Glen, John, 98
Glen Sannox coach tour, 128
Gourock extension, 22-3, 31-2, 34, 39, 41-3, 46-7
Grampian trains (Cal Ry), 203
Greenfield, J. W., 148
Greenock & Ayshire Ry, 21
Greenock & Wemyss Bay Ry, 19, 43, 52, 55, 97, 189
Greenock, Burgh of, 74
Greenock Harbour Trust, 74
Grey, Sir Edward, 211
Gregor, Capt, 174
Grierson, Henry, 141
Guthrie, Miss, 87

Haliday, Geo Ltd, 202
Hamilton, Lady Mary, 170, 174
Hargreave, Rev Alfred, 138
Hastings, St Leonards & Eastbourne Steamboat Co, 134
Hawthorn, Leslie & Co, 154-5
Haythorn Tubulous Boiler Syndicate, Ltd, 142, 145-6
Hayward Family, 229
Henderson, D. & W., 26, 103-4, 225-6
Henderson, John, 104
Highland Light Infantry, 225
Hill & Co, 31
Hillhouse, Percy, 191
Hubner, Mr, 138
Hunter, Graeme, 137
Hutson & Corbett, 27-8, 31, 101, 106, 147, 266

Iff, Herr, 225
Inglis, A. & J., 30, 48, 110, 120, 131, 141-2, 146-8, 150, 158, 185, 188, 194, 207-8, 241
Inglis, Dr John, 131, 208
Innellan Pier, 231
Institution of Naval Architects, 159

Jubilee Review at Spithead, 154

Kaiser, The, 183, 211-2, 228
Keith, Capt Angus, 181
Kerr & Rayner's feed heater, 123
Kilmun trade, 31, 33, 37
King, Wm, & Co, 82, 110

Lamont, Mr, 55
Lanarkshire & Ayrshire Ry, 46-7
Lanarkshire coal traffic, 47
Larne & Stranraer service, 176
Lee, Capt, 184
Lennox, William, 204-5

Leopoldina Railway Co, 187
Letters to the Editor—Campbeltown service, 112-14; Dunoon service, 123; GSWR train services, 83, 91-2; interests of the working classes, 70, 72; Kilchattan Bay service, 226; Millport affair, 196, 198; music on board, 231; oil fuel, 221; pier dues, 232-3; sewage problem, 218-19; smoke problem, 220; steamer racing, 79-80, 124; Sunday religious services on *Victoria*, 138; train speeds on GSWR, 93
Leyland, Capt Christopher, 165, 178, 264-5
Leyland, Miss Dorothy, 165
Lipton, Sir Thomas, 226
Lloyd George, Rt Hon David, 203
Lock Eck tour, 184
Loch Lomond steamers, 76, 194
Lochgoil & Inveraray Steamboat Co Ltd, 206, 210, 286
Lochgoil & Lochlong Steamboat Co Ltd, 25, 74, 101-2, 206, 221, 245, 255, 285-6
Lochlong & Lochlomond Steamboat Co, 25-6, 101, 184
Locomotive coal costs, 236
Locomotives—Caledonian Ry: *Cardean* class, 203; Drummond 0-6-0, 189; *Dunalastairs*, 189; *Eglinton*, 51; 'Greenock Bogies', 43, 189; McIntosh 812 class 0-6-0, 189; McIntosh 908 class 4-6-0, 203
Locomotives—GSWR: 'Greenock Bogies', 83; Manson No 8 class 4-4-0, 92-3; Smellie 'Wee Bogies', 82-3, 92
Logan, Mr, 84
London & Glasgow Shipbuilding Co Ltd, 30

Maguire, Joseph, 217
Manchester Ship Canal, 116
Manson, James, 21, 83, 92
Marquis of Breadalbane, 41, 47, 59, 121
Marquis of Bute, 195, 200
Marquis of Graham, 170, 174
Marquis of Lorne, 60
Martin, Thos, 130
Melville, Mr, 92
Metropolitan Railway Carriage & Wagon Co, 82
Midland Ry, 184
Millport 'siege', 183, 194-203
Millport Town Council, 194-6, 199-202
Millport Visitors' Club, 63
Minto, Mr, 103
Montgomerie Pier, Ardrossan, 51
Morrison Capt Robert, 52, 64-5, 172. 213
Morton, Robert, 48-9, 60, 77, 86-8, 130
Morton & Williamson, 49, 77
Muir, Mr, 223
Murdoch & Murray, Ltd, 210
Murray, Mr, 74
Music on board, 228-31
McArthur, J. & Co, 101
MacBrayne, David, and David MacBrayne, Ltd, 23, 37, 40, 68, 73, 95, 106, 115, 151, 176, 188-9, 219, 237, 244, 251-2, 255, 258, 285-6
McCaig, John Stuart, 39
McCallum, Capt Hugh, 141
Macdougal, Capt Allan, 172
McGregor, Capt Colin, 81, 214
McIntosh, John F., 189, 203
McIntyre, Hugh, 31, 111
McKechnie, Capt John, 204-5
McKellar, Capt A., 172, *197*
McKinlay, Capt Dan, 103, 120, 132
McKnight & Co, 38, 101, 227
MacLachlan, Capt Archd, 98

MacLachlan, Capt John, 136
McLaren, Thos, 86
Maclean, Miss, 104
Maclean, William, 104
McMillan, Eben, R., 214
McNeill, Capt Duncan, 120
McPhail, Capt, 149-50
McPhedran, Capt D., 123

Napier, David, 244
Napier & Miller, Ltd, 207
Napier, Shanks & Bell, 77
National Bank of Scotland, 156, 164
Naval Architecture, Chair of (University of Glasgow), 191
North British Ry, 20-1, 27, 40, 43, 57-8, 84-5, 101, 119, 133, 147, 184, 188, 192-4, 206, 217, 243, 247, 251-2, 254, 259, 271, 285-6
North British Steam Packet Co, 20-1, 27-8, 39, 94-6, 101, 103, 110-11, 116, 118-19, 131-2, 139, 141, 147-8, 150, 152, 188, 193, 228, 239, 254, 266-7, 286

Oil fuel, 221-2

Paddleboxes, 251-8
Paisley sailings, 99, 106, 109
Parson, Hon Charles, 153-5, 158-9, 163-5, 182, 265
Parsons Marine Steam Turbine Co Ltd, 154-6, 163-5, 168, 173, 182, 248, 262-5
Parsons, Mrs, 158
Paterson, Alexander, 40, 73
Pember, Mr, QC, 40
Pier dues, 231-3
Prince and Princess of Wales, 51, 121, 225
Prince's Pier rebuilding, 85, 91-2
Princess Louise, Marchioness of Lorne, 59
Princess Louise, Duchess of Fife, 191
Punch, 222

Queen Alexandra, 121
Queen Victoria, 60, 64, 152

Rankin & Blackmore, 44-5, 48, 56-7, 59, 61, 65, 185, 210
Rathlin Island & Killough services, 121
Reid, John, & Co, 37, 44, 56
Renshaw, Sir Charles Bine, 203
Robertson & Co, 30
Robinson, Heath, 248
Rowan, D. & Son, 77, 91
Rowatt, Provost, 200
Royal Clyde Yacht Club, 62, 225
Royal Mail, 24-5, 73, 75, 285
Russell & Co, 59-60, 124, 134, 193, 252

Scott & Co, 48, 59, 158, 166
Scott, Sir Walter, 254
Scottish Excursion Steamer Co Ltd, 111, 114-15, 127, 286
Seath, T. & Co, 27, 110
Second class travel abolished, 43
Sewage problem, 217-9
Ship Canal Passenger Steamer Co (1893), Ltd, 116
Ships—see separate index
Sinclair, William, 199
Smellie. Hugh, 21, 82-3, 92
Smith, Lt A. W. B., RN, 163

Smoke fiend, 219-22
Smyth's propellers, 167
Somerville, Cameron, 229
Southampton Shipbuilding & Engineering Co, 127
Staffa and Iona service, 25
Steam Vessels Committee (GSWR), 75, 77, 85, 172
Stewart, Rev Alex, *197*
Storms, 223-4
Sunday sailings, 135-9

Tay Bridge, 22
Temperance, 26-7, 134
Thames steamers, 49, 100, 252
Thompson, Sir James, 34, 39-40, 168
Thomson, J. & G., 23, 77, 80, 86, 120-1, 130, 168, 186, 241
Thomson, Mr, 226
Thevenet, M Jean, 86
Ticket agreements, 85, 123
Trades House of Glasgow, 94
Trains, new, for coast services: Caledonian Ry, 203; GSWR, 82; North British Ry, 97
Traffic decline, GSWR, 67-8
Turbine engines, 153-4
Turbine Steamers, Ltd, 164, 175, 177, 181, 210-11, 256-7, 259, 285-6
Turbine Syndicate, 156, 160, 164, 257, 262-4, 285-6
'Turkish Fleet', 29, 68, 75, 95, 109, 158, 256, 285-6

Uniforms, 252-3
Union Castle, Co, 77, 254
Union Steamship Co (New Zealand), 204

Watson, George Lennox, 225
Watson, Sir W. Renny, 84, 115
Wemyss Bay rebuilding, 189-90
West Coast route, 184
West Highland Ry, 120
Williamson, Capt Alexander (Sr), 21, 26, 29, 32, 68, 75, 88, 109-10, 134, 205, 256, 286
Williamson, Capt Alexander (Jr), 75-6, 92, 94, 109, 139-40, 172, 174, 187, 195
Williamson, Capt James, 26, 37, 41, 44, 46, 48-9, 55, 57, 60, 63, 65, 73, 75, 94, 110, 120, 133-5, 142, 145, 168-70, 185-6, 191, 195-6, 221, 234, 237, 242, 252
Williamson, Capt John, 88, 109, 115, 124, 130-1, 134-5, 146, 156, 158, 164, 169, 175, 177, 181-2, 192-3, 205, 210, 249, 256, 259, 262, 264-5, 270-2, 285-6
Williamson, Miss Maud, 57
Wilson, Aird, 112
Wilson, John, MP, 70, 73
Winton Pier rebuilding, 81
Wire mesh on rails, 217, 267
Wright, Whitaker, 228

Yachting, 224-8
Yachts—see separate index
Yarrow boilers, 154-5, 157

INDEX OF STEAMERS AND OTHER VESSELS

CLYDE STEAMERS

Adela, 56, 98-9, *144*, *(277)*
Ardmore (ex Sultan), 115
Argyle, *(277)*
Argyll, 113
Atalanta, *108*, 173-4, 182, 205, 247, 258, *(277)*
Athole, *(277)*

Balmoral (ex Lady Brisbane), 29, 96, 106, *(277)*
Benmore, 29-30, 134, 192, 222-3, 259, *(277)*
Britannia—see Vulcan

Caledonia (built 1889), 41, 44-5, 56, 60, 62, 65, *161*, 192, 221-3, 237, 239, 242, 252-3, 258, 274-5, *(277)*
Caledonia (built 1934), 49
Carrick Castle—see Culzean Castle
Chancellor, 25-6, 67, 75-7, 82, 101, 185, 187, *(277)*
Columba, *17*, 23-5, 40, 42, 44, 46, 50, 66, 73, 95-6, 100, 106, 117, 151, 166, 176, *180*, 184, 189, *197*, 219-20, 244, 252, 255, 259, *(277)*
Culzean Castle (ex Windsor Castle), 127-8, 146, 255, 258, 286, *(277)*
Cumbrae (ex Victory), 31, *(277)*

Dandie Dinmont, 120, 259, *(277)*
Daniel Adamson—see Shandon
Davaar, *18*, 30, 117-18, 128-9, 192, 205, 245, 258, 270, *(278)*
Diana Vernon, 28, 96, 118, 120, 184, *(278)*
Duchess of Argyll, *108*, 170-2, 175-8, 181, 206, 248, 253, 258, *(278)*
Duchess of Fife, 186, 190-1, 210, *216*, 237, 241-3, 249, 253, 258, *(278)*
Duchess of Hamilton, 47-52, 55-6, 58, 60-4, 68, 79-80, 121-2, 127, 153, 157, 159, 172, 185, 213-14, 225, 232-3, 239, 243, 252-3, 258, 274-5, *(278)*
Duchess of Montrose, 179, *180*, 185-8, 191, 241-2, 245, 249, 258, *(278)*
Duchess of Rothesay, *35*, 121-4, 130, 134, 158, 172, 186, 191, 241-2, 247, 249, 258, *(278)*
Duchess of York (ex Jeanie Deans), 146, 193, 228-30, 258, 285-6

Eagle, 116, 214, *(278)*
Eagle III, *126*, 207-10, 258, *(278)*
Edinburgh Castle, 25, 67, 186, 206, 210-11, 221, 223, 245, 252, 256, 286 *(278)*
Elaine, *(279)*

Galatea, 41, 45, 62-3, 93, 168-9, 185, *197*, 225, 239, 274-5, *(279)*

Gareloch, 28, 96, 101, 103, *(279)*
Glen Rosa, 86-8, 115, 193, 258, *(279)*
Glen Sannox, frontispiece, 80-1, 84, 88, 94, 124, 130, 148, 153, 172-3, 175, 214, 232, 241, 243-4, 258, *(279)*
Glenmore, 124, 127, 134, *(279)*
Grenadier, *18*, 24, 159, 188-9, 224, 244, 255, 259, *(279)*
Guinevere, 30, 106, 110, *(279)*
Guy Mannering (ex Sheila), 28, 97, 102, 110-11, 116, 118-19, 251, *(279)*

Herald, 113, *(279)*

Iona, 24-5, 40, 42, 106, 151, 166, 176, 189, 214, *215*, 219, 224, 228, 244, 252, 258, *(279)*
Isle of Arran, *54*, 110, 119, 127-8, 192, 207, 228, 256, 258, *(279)*
Isle of Bute (ex Guy Mannering), 116-17, 119, 193, 210, 228
Isle of Cumbrae (ex Duchess of York), 193, 258
Isle of Skye (ex Madge Wildfire), 211, 258
Ivanhoe, 26-7, *36*, 37, 41-2, 44-5, 62-3, 67, 73, 104, 115, 133-4, 185, 206, 209, 245, 252-3, 257-9, 285-6, *(280)*

Jeanie Deans, 28, 95-6, 102, 111, 116, 118-20, 131, 146, 258, 267, *(280)*
Juno, 140, 148, 152, 228, 241, 247, 254, 258, *(280)*
Jupiter, *89*, 130, 132-4, 139-40, 152, 160, *179*, 188, 241, 247, 258, *(280)*

Kenilworth, 141, 147, 207, 259, *(280)*
King Edward, *90*, 107, 157-60, 163-7, 169-71, 173, 175-7, 181-2, 184-5, 206, 217, 247-8, 257, 259, *(280)*
Kinloch, 30, 205, 245, 258, *(280)*
Kintyre, 30, 112-13, 117, 129, 183, 204-5, 223, 245, *(280)*
Kylemore (ex Vulcan), 134, 193, 205, 249, 259

Lady Brisbane—see Balmoral
Lady Clare, 103, 118, 120, 193, *(280)*
Lady Rowena, *71*, 102-3, 118, 120, 188, 194, 210, 254, 258, 266-70, *(281)*
Lancelot, 99, *(281)*
Lord of the Isles (I), 23, 25-6, 42, 44, 46, 66-7, 95, 99, 100-1, 104-5, *144*, 192, 245, 252, 258, *(281)*
Lord of the Isles (II), *53*, 103-6, 117, *143*, 166, 184, 206, 210, 220, 245, 252, 255-,6 259, 286, *(281)*
Lucy Ashton, 27-8, *71*, 118, 120, 150-1, 188, 243, 259, *(281)*

295

Madge Wildfire, 31-3, 37, 42, 57, 62, 65, *162*, 192, 209-11, 252, 258, 274-5, *(281)*
Marchioness of Breadalbane, 57, 123, 185, 201, 205, 252, 258, 274-5, *(281)*
Marchioness of Bute, 57, 62, 176, 185, 201, 205, 274-5, *(281)*
Marchioness of Graham—see Duchess of Argyll
Marchioness of Lorne, 49, 59-61, 63, *72*, 122, 142, 186, 214, 241-2, 252-3, 258, 274-5, *(281)*
Marmion, *125*, 193-4, 243, 259, *(281)*
Marquis of Bute, 75-6, 82, 109, 193, 256, *(282)*
Mars, 172, 187, 193, 242, 245, 249, 258, 284, *(282)*
Meg Merrilies, 28, 31, 33, 37, 41-2, 62, 65, 121, 142, 145-6, 187, 189-90, 223, 239, 274-5, *(282)*
Mercury, 78, 81, 87-8, 217, 224, 243, 249, 258, *(282)*
Minerva, 86-7, 93, 115, 193, 258, 271, *(282)*

Neptune, *54*, 78-80, 87-8, 92, 129, 140, *162*, 243, 248, 258, 261, *(282)*

Queen Alexandra (I), *107*, 158, 164-9, 171-2 175-8, 182, 222, 257, *(282)*
Queen Alexandra (II), 178, 181, 257, 259, *(282)*
Queen Empress, 210, *216*, 257, 259, *(282)*

Redgauntlet, 119-20, 131-2, 141-2, 147-50, 206, *(282)*

Scotia, 29-30, 52, 68, 74, 76, 79, 86, 88, 213-14, 243, 245, 256, *(282)*
Shandon, 106, 109, 116, *(282)*
Strathmore, *126*, 134, 146, 156, 205, 249, 271, *(282)*
Sultan, 75-6, 88, 109-10, 115, *(283)*
Sultana, 75-6, 109, 115, 130-1, 134, 256, *(283)*

Talisman, 131-2, 141, 147, 207, *216*, 225, 259, *(283)*

Vulcan, 245
Vulcan (ex Britannia), 193, 201, 205, *(283)*
Viceroy, 29, *72*, 75-7, 109, 141, *180*, 205, 256, *(283)*
Victoria, 29, *36*, 52, 95, 97-9, 111, 114-15, 118, 127, 135-9, 146, 225, 239, 257-8, *(283)*
Vivid, 93, 106, *(283)*

Waverley, 147-8, 150, 152, 194, 228, 241, 243, 259, *(283)*
Windsor Castle, 25, 67, 102, *(283)*

MISCELLANEOUS SHIPS

(1) COASTAL PASSENGER

Albion (ex Slieve Donard), 87
Britannia, 38, *215*, 227-8
Cobra, 46-7
Dromedary, 129
Fairy, 64
Fairy Queen, 184
Gairlochy (ex Ardmore), 115
Koh-i-Noor, 190
Lorna Doone, 77-8
La Marguerite, 190
Princess Maud, 176
Slieve Bearnagh, 121, 130
Slieve Donard, 87, 129
Waverley (P. & A. Campbell), 31, 33, 37
Westward Ho!, 38, 147, 236
Wemyss Castle (ex Gareloch), 101-2

(2) NAVAL SHIPS

HMS Cobra, 154-5, 157, 163
HMS Crane, 163
HMS Viper, 154-5, 157, 160
HMS Vulture, 163

(3) CUNARD SHIPS

Caronia, 173
Carmania, 173
Lusitania, 173
Mauretania, 173

(4) MISCELLANEOUS STEAMSHIPS

Craigielee, 202
Emily, 129
Elizabeth, 202
Godmunding, 141
Maori, 204
Osprey, 57
Ranger, 129
Turbinia, 154

(5) YACHTS

Ailsa, 226
Britannia, 226
Kariad, 228
Meteor, 228
Satanita, 226
The Shamrocks, 226, 228
Sybarita, 228
Valkyrie, *198*
Valkyrie II, 225-6
Valkyrie III, *198*, 226